D0915462

BEEF PRODUCTION

SCIENCE and ECONOMICS

APPLICATION and REALITY

AUTHOR

DAVID P. PRICE Ph.D.
Private Consultant
Las Cruces, NM

Contributors

THOMAS D. PRICE M.S.
Interwest Ranch Management Inc.
Pendleton, OR

TOMMY BEALL
Director Economic Research,
Cattle Fax
National Cattlemen's Assoc.
Denver, CO

Editorial Consultant

D.C. CHURCH Ph.D.

Corvallis, Ore.

Copyright © in 1990 by D.P. Price. All rights reserved. No portion of this book may be reproduced in any manner without written permission of the publisher.

First Printing 1981
Second Printing 1985
Third Printing 1990

Published and Distributed by
SWI Publishing
P.O. Drawer 3 A&M
University Park, New Mexico 88003
United States of America

Telephone 505-525-1370
Fax 505-525-1394

Library of Congress Catalog No. 81-51944
ISBN 0-9606246-0-0 Hardcover
ISBN 0-9606246-3-5 Softcover

SPECIAL ACKNOWLEDGMENT

Professor John K. Riggs
Texas A&M University
Distinguished Scientist, Respected Beef Cattle Authority,
Impeccable Gentleman

The study of science is essentially the study of the ideas and experiences of others. Indeed, very few of the concepts presented in this text are my own. As the author, I have simply tried to record and organize the teachings of a myriad of people who have been so kind as to share their knowledge with me.

Of these people, Professor Emeritus John Riggs has had the greatest influence. Many of the concepts developed in this text are his. His greatest contribution, however, has not been in the form of discrete facts or information. Well versed in the basic sciences affecting agriculture, Prof. Riggs has the gift of being able to draw information from the basic sciences and combine it in such a matter so as to solve or explain real life production problems. As a graduate student, I recognized that ability, and was motivated by it. In essence, he taught me the true value of knowledge. In keeping with his training, I have tried to write this textbook around that basic philosophy, and do sincerely hope that I have been successful.

DEDICATED

to my wife

Cheryl

for giving up a promising career in marine biology
in order to allow me to pursue my profession as a
beef cattle consultant

PREFACE

The purpose of this book has been to present academic and scientific principles in such a manner that they can be readily understood in their application. The ultimate purpose has been to supply practical, useful information to those wanting to learn to manage beef cattle operations efficiently; the book being aimed primarily at undergraduate students and producers actually in the business.

In addition, the chapters have been arranged in such a manner that the book may also be used as a reference . . . that the reader may "grow" with the book as his knowledge of Animal Science increases. For example, there are separate sections on the different types of beef production: cow-calf, stocker operations, cattle feeding, etc. In those chapters, management procedure explanations are kept as basic as possible. For more detailed scientific explanations of individual points, the reader is referred back to other chapters dealing strictly with science; e.g. nutrition, genetics, reproduction, etc. In that way, a beginning student or a producer can get quick, useful management information, even though he may not yet have the technical/formal training in the sciences involved. The science chapters are there, however, so the reader can obtain more detailed information, and indeed, grow with the book.

Beef production, as it exists today, is an extremely competitive business in which miscalculations of only a few percentage points can be financially destructive. For that reason, cattle and feed measurements have been described in English rather than metric units. That is, English measurements are used in U.S. agriculture today, and will continue to be used for a long time, regardless of what the scientific community may do. To teach students who would go out and manage cattle operations to think of cattle in terms of kilograms, is doing them an injustice, in the opinion of the author. Conversion of kilograms to pounds is quite simple. But one simple arithmetic error in a sales ring, commodity office, or on the telephone late at night (while physically or emotionally tired) can cost thousands of dollars. Therefore the author has deemed it necessary to use the English system for the purpose of this book.

TABLE OF CONTENTS

PART I — PRODUCTION AND MANAGEMENT

APPENDICES

STATISTICAL DATA

(Courtesy Texas Cattle Feeder's Association)

CHAPTER 1   INTRODUCTION

In recent years beef consumption/production has been criti- cized for several reasons by urban groups. Although well inten- tioned, this criticism has come as a result of incomplete information. The truth of the matter is that beef is a healthful and nutritious source of protein. Because of the ruminant's ability to utilize roughages and non-protein nitrogen, beef production will probably become of greater importance as population demands decrease the availability of grains and natural proteins for livestock pro- duction. In addition, if Japan and the Soviet Union do not cur- tail their relentless and irresponsible overfishing of the world's oceans, conventional agriculture may also be required to supply a portion of the world's protein that is currently being supplied by fish.* This, obviously would place an even greater importance upon beef production.

THE FEEDING OF GRAIN TO LIVESTOCK

The most recent criticism of beef consumption/production by urban societies has been that it is an inefficient use of grain. The basis of this thought being that feedlot cattle convert grain into a liveweight gain at a ratio of about 7.5 : 1, whereas swine convert at about 3.5 : 1 and poultry at about 2 : 1.

The fallacy in this idea is that only about 40% of the final slaughter weight of a steer is attributable to grain feeding. The typical feedlot steer doesn't come into the feedlot until it weighs about 650 lb. If the amount of grain fed during that period is divided by the animal's final slaughter weight (about 1050 lb), a grain conversion of about 3 :1 is obtained. In addition, breeding herds of cattle receive no grain (held on pasture), where as swine and poultry breeding stock must be fed grain diets.

But what gives the cattle business such an advantage over the other livestock enterprises for the future is the fact that cattle don't have to be fed any grain at all. At the present time feedlot cattle are fed grain rations primarily as a matter of economics. As mentioned earlier, cattle convert grains at a ratio of about

---

* Japan refuses to abide by World Fishery Conservation recommendations. The Soviet Union signs conservation treaties, etc., but then proceeds to take mul- tiples of her assigned quotas. As a result, extensive damage is being done to fish populations around the world. (McCloskey, 1979).

7.5 : 1, whereas roughages are usually converted at a ratio of about 12.5 : 1. At 6¢/lb for grains and 4.5¢/lb for roughages, the grain will produce a cost of gain of about 45.0¢/lb vs. 56.25¢/lb for the roughages. In addition, cattle fed grain rations gain faster and are therefore ready for slaughter earlier, which reduces the daily management expense.

Along with criticism of grain feeding there has been a lot of talk about grass fattening of beef. In the U.S. grass fattening of beef is not logistically possible. There just isn't enough grass to do that, as the area that would be analogous to the grasslands of Argentina or Australia has long since been put to the plow; i.e. the Great Plains of the Midwest. To go back to grass fattening would greatly reduce overall agricultural production.

However, it is the ability of cattle (ruminants) to utilize grass and grasslike roughage materials that will ensure that the cattle business remains a viable enterprise. Indeed, as more of our grains must be diverted for human consumption, the dependence upon beef as a protein source should increase. Just how much will depend upon economics.

There are vast quantities of roughage and roughage-like by-products (cornstalks, straws, sawdust, pulp residues, etc.) that can be used as cattle feedstuffs. The technology for using these products (treatment with NaOH, ammoniation, etc.) is available, the only drawback being the cost. At the present time it is not economically feasible to treat and use these products. The point, however, is that the potential for using a vast array of "waste" products is available if it is ever needed. If all grains and vegetable protein products had to be diverted for human use, beef production would become extremely important since residues from crops used for human consumption could be fed to cattle. Swine and poultry production would be eliminated.

The author doubts, that there will ever be a situation where all grains and vegetable proteins are used for human consumption. What will more likely occur is that grains and vegetable proteins will become relatively expensive, which will greatly curtail their use as livestock feed. This, of course, will proportionately increase the cost of swine and poultry; and as the cost increases, and the technology for feeding residues to cattle becomes more refined, more emphasis will probably be placed on beef as an animal protein source.

BEEF AND THE CHOLESTEROL SCARE

Arteriosclerosis is a disease created by cholesterol/fatty acid

Figure 1-1a. Commercial steam hydrolyzation plant for chemical processing of low quality roughages. Cotton gin trash, corn stalks, wheat straw, etc. are placed in the two spherical "digesters" and treated with nitric acid under steam pressure. The process breaks down much of the fiber and apparently forms complex sugars, as the resultant product carries a sweet but acrid aroma.

Ruminants, of course, are the only animals that can utilize such products. Thus, if population demands decrease the availability of grains for livestock feeding, it is operations such as this one that will ensure the viability of the cattle feeding industry. (Howard Dedrichson, Dumas, TX. Owner-operator.)

Figure 1-1b. Cotton gin-trash before (left), and after (right) treatment.

deposits in the arteries which impede or block the movement of blood and cause subsequent heart failure. As cattlemen are all too weary of hearing, the consumption of beef (meat) is often said to contribute to arteriosclerosis.

The problem is that there is very little known about the biochemistry of fat digestion and absorption in humans. In studies with animals, it has been found that different species vary greatly in their ability to utilize cholesterol. Rabbits and chickens absorb cholesterol directly and thus high levels of dietary cholesterol elevate blood cholesterol accordingly (Groen, 1952; DHEW, 1971). However, in rats and dogs increased dietary cholesterol has not increased blood cholesterol. In order to get increased cholesterol uptake in rats and dogs, thyroid activity must be depressed (DHEW, 1971). In studies with humans having already high blood cholesterol levels, reduced cholesterol consumption has been shown to reduce blood cholesterol levels. But in studies with normal humans, diet seems to have little effect on blood cholesterol levels. (Kannel et al, 1971; Kahn et al, 1969)

In normal people cholesterol is apparently broken down into intermediary compounds rather than absorbed directly. It would seem that humans who have abnormally high blood cholesterol due to cholesterol in the diet, have some sort of biochemical/digestion dysfunction such as the depressed thyroid activity demonstrated with rats and dogs. But the general public has been given the impression that eating animal fats increases the likelihood of heart disease for everyone. This has been capitalized upon by the prepared food industry in marketing such products as textured vegetable protein "meats", synthetic eggs, milk and margarine, etc. Advertisements for these products try to leave the impression that it is healthier to consume these products rather than real meat, eggs, butter, etc.

In some cases the advertising has been untruthful and misleading. The most flagrant being the margarine advertisements that typically refer to their products as being made with unsaturated or "soft" fats; i.e. superior to the saturated "hard" fats found in butter. While the safflower, corn oil, etc. that is used in the manufacture of margarine is unsaturated in the beginning . . . it must be saturated or hardened in order to be made into a stick.

The consumer may feel he is better off consuming these products, but they may actually be more injurious to his health than the real thing. In studies done with primates, it has been found that certain unsaturated fats increase cholesterol absorption more than when saturated fats are in the diet (Tanaka, 1977). Furthermore it is the author's undocumented opinion that even overlooking the possibility that the oils used in these products may increase cholesterol absorption, synthetic products probably create a greater overall health hazard due to the liberal use of artificial coloring and flavoring factors, and antioxidants.

Recently there have been studies which suggest that refined carbohydrates (sugar in particular) may contribute to arteriosclerosis (Yudkin, 1968). Sugar is broken down very quickly and goes into the bloodstream in the form of glucose in a matter of minutes. This can be dramatically exemplified by diabetics who can be trembling (due to lack of glucose), and then be perfectly normal within 5 or 10 minutes after ingesting a small amount of sugar. Fats, on the other hand, are digested over a period of several hours (this is why particularly greasy foods are difficult to digest).

When a relatively large amount of sugar is eaten and subsequently broken down, the glucose level in the bloodstream is increased quite rapidly. However, homeostasis will not allow the glucose to accumulate in the blood very long. It is pulled out of the bloodstream and synthesized into various other compounds; e.g., glycogen, triglycerides, and possibly cholesterol.

Research work with unsaturated fats and refined carbohydrates as contributing factors to arteriosclerosis has been rather limited, and so drawing conclusions is premature at this time. However, in the author's opinion, it seems more plausible to assume that refined vegetable oils and carbohydrates are more prone to elevate blood cholesterol levels than animal fats. The author is of this opinion because of the fact that elevated serum cholesterol levels are rarely found in primitive societies (DHEW, 1971). Most primitive cultures consume substantial amounts of fat, and very little refined vegetable oil or carbohydrates. Elevated serum cholesterol and arteriosclerosis is a disease associated with affluent societies, that eat large amounts of sugar, flour, and processed vegetable oils. Yudkin (1968) has discussed studies with the Masai and Samburu tribes of East Africa that have a diet that consists primarily of milk, blood, and meat. Although their saturated fat intake is higher than any group of people for whom international dietary statistics are available, their experience with coronary disease is low.

The variable that makes it difficult to draw conclusions on diet alone, is the vast difference in physical activity between primitive and affluent societies. Primitive peoples are much more active and exposed to more environmental physical stress than civilized man, and certainly this is protective of heart disease. Even so, the author still feels that with the large amounts of animal fats consumed by these peoples, it is reasonable to draw the conclusion that animal fats are not necessarily a predisposing factor to arteriosclerosis.

In summary, the author sincerely feels that physiologically normal* human beings consuming a well balanced diet, utilizing animal proteins, vegetables, and a minimum amount of refined carbohydrates and vegetable oils, coupled with daily excercise, should not have undue fear of heart disease. In essence, the author feels that beef is indeed a healthful and nutritious protein source that has been maligned through sensationalist popular press articles, advertisements, and self-appointed experts with no real training in nutrition.

THE RESIDUE SCARE

Though not as serious as the cholesterol issue, beef has also received a lot of bad publicity over the use of feed additives and implants. When the FDA** has raised issues concerning the use of these compounds, cattlemen's groups, the drug companies, and feed manufacturers have fought vigorously to keep the products. As a result of these "fights", beef has received a lot of bad publicity.

Because of this publicity, many consumers believe that when they consume beef they also ingest the same steroids, antibiotics, etc. that cattle are fed. This is not true. The FDA will not allow the sale or use of any drug that leaves a residue in the edible portion of livestock. Each drug that is used comes with a specified withdrawal time. Withdrawal time referring to length of time that the cattle must cease receiving the drug before slaughter, so that no trace of the drug remains in the edible portions of the meat . . . and the vast majority of feedlots are very conscientious about observing withdrawal times.

As a general rule feedlots obey withdrawal times not out of any love for the FDA, or as a matter of public concern . . . but simply because they don't want any residues found in the carcasses of their animals. If residues are found, the entire carcass is condemned. With carcasses worth anywhere from $500 to $1,000 apiece, condemnation is a very expensive proposition. Aside from the condemnations the feedlot is also quarantined until the remaining animals are proven to be free of tissue residues

---

* Persons with abnormal cholesterol metabolism may need to consume low cholesterol diets.

** U.S. Food and Drug Administration

of the compounds involved. This can be very costly to the feedlot for two reasons: (1.) When cattle are ready to go to market they must be sold within a few days or gain costs increase sharply. (2.) If cattle are held on feed past their optimum market weight, they become overly fat, which can substantially reduce the price a packer will pay for them. With several hundred or thousand cattle on feed, it just doesn't make sense to take those kinds of risks in order to get the very minimal advantages afforded by extended use of drugs.

In addition to what many cattlemen might term an almost hysterical concern over residues, the FDA will also not allow the use of any drug proved to be harmful . . . regardless of whether or not residues occur. In the case of possible carcinogens, if the compound causes cancer in laboratory animals at any concentration, it is banned. In most cases these products are continuously fed to laboratory animals at thousands and even millions of times the concentration a human could ever get from an occasional piece of meat containing residue. The important point here is that a drug can be banned if it is harmful (at any concentration), even if residues are not found, or if it is not found to be harmful, but leaves a residue. Cattlemen and drug companies can argue about the validity of such extraordinary precautions, but to the consumer, it is his assurance that the meat he purchases is indeed a wholesome and healthful product.

EXPORT POTENTIAL

As a final note the author would like to point out the possible future export potential for beef.

It is well established that beef is the most preferred source of protein by most societies, and as the wealth of nations increases, so does the demand for beef. However, at the present time there is relatively little beef exported from the U.S. The underdeveloped nations don't have the wherewithal to pay, and the developed nations typically have restrictive trade policies. In Japan and Europe, retail beef sells for several times as much as beef does in the U.S.

Farmer and commodity associations are becoming more and more aggressive in seeking export markets. This, coupled with the obvious overseas demand for beef (people willing to pay up to $20/lb), would make it seem logical that at some point in time, there will be a large export market for U.S. beef. However, trade policies, like all politics, do not always follow the dictates of logic.

There is a situation developing, that in the author's opinion may override the politics of world trade and force the opening of barriers for at least one major country . . . Japan. At the present time fish is the main source of protein in the Japanese diet. As pointed out in the introductory paragraph to the chapter, Japan (and the Soviet Union) is causing serious commercial damage to a number of fish populations around the world. This kind of irresponsible behavior cannot be justified out of necessity (to feed a starving people), as Japan has one of the most restrictive meat import policies in the world. A clique of Japanese aristocrats have had absolute control over meat imports since WWII, and they allow only a token amount to come into the country.*

The author does not mean to imply that the U.S. and other countries could export enough meat to replace the fish in the Japanese diet. Rather, it is the author's contention that enough meat could be exported to allow the Japanese to follow responsible fishing practices and quotas.

If Japan continues to disregard World Fisheries Conservation guidelines, it seems logical to assume that at some point in time she will be forced to open up her restrictions on meat imports to replace decreased fish catches.** That is, Japan is not physically capable of significantly expanding her conventional agriculture. Such a situation would also adversely affect the Soviet Union. Unlike Japan, the USSR is logistically capable of expanding her conventional agriculture (meat production), the only thing holding her back being the inherent inefficiency of the communist form of agriculture.

CONCLUSIONS

It is the author's opinion that the long term future for beef production is rather bright. This thought is based on the fact that beef is a wholesome and nutritious source of protein which should increase in importance as the world's supply of grain and vegetable proteins becomes tighter.

---

* In 1978, after much negotiation, Japan agreed to increase beef importation from 16,800 to 30,800 metric tons by 1983. While this may seem to be a substantial increase, the total amount is still only equal to about 108,000 slaughter steers - approximately the yearly production from just one large feedlot.

** At the sake of belaboring the point, the author feels it necessary that every agriculturist (particularly those involved in academic or government work) be aware of this situation. Agriculture and fisheries are often separated but they are actually one and the same, the production of food (protein). There has been much talk of turning to the sea, but to date there has been much more technology and investment applied to taking from the sea than replenishing the sea.

However, in the interim period, we must deal with domestic demand. As discussed in detail, beef has been maligned. While unfounded, the general public is not aware that the information they have been provided is incomplete. All they know is what our sensationalist popular press has told them.

If we are to maintain our domestic demand for beef, we must make complete information available to the public. As cattlemen, this means we are going to have to do two things:

(1.) Help finance and otherwise initiate research in human nutrition utilizing normal individuals (rather than medical patients with existing cardiovascular problems, etc.).

(2.) Disseminate the information in a form that will actually reach the consuming public.

The Texas Cattle Feeders Association has taken the initiative and at the time of writing had helped finance two cholesterol studies (one at Texas A&M, and one at Baylor Medical School), and (along with the National Cattleman's Association) has been purchasing full page advertisements in popular magazines describing the nutritive value of beef. This second point is very important . . . we can make the information available, but we cannot expect the popular press to disseminate it. We must understand that the popular press sells its product with sensationalism. "Dietary Cholesterol Is Broken Down into Intermediary Compounds", certainly doesn't sell a magazine, newspaper, or TV program like "Heart Attacks and Cancer Linked to Meat Consumption". If we want the public to know the truth, we are going to have to tell it to them ourselves.

In summary, if we can educate our consuming public to the nutritive value of beef, we should be able to maintain a good domestic demand. If we can get our politiians to negociate trade policies that will remove unfair restrictions on beef, we could enjoy a greatly increased foreign market. Moreover, we have an excellent product and the future for our industry looks bright. We have logic and scientific fact on our side. Unfortunately, however, we cannot expect our product to stand on its merits. To ensure our future, we must become increasingly involved in marketing.

CHAPTER 2   WESTERN COW-CALF RANCHING

RANGE MANAGEMENT

Figure 2-1. Unfortunate example of excessive grazing animal impact. Note the erosion beginning to occur just under the crest of the ridge. Courtesy M.M. Khotmann, Texas A&M Univ.

Range management was chosen as the first topic of this chapter as it is the most fundamental and important aspect of ranching. Far too many ranchers lose sight of this fact. It must be realized and always borne in mind that grass and forage are the commodities that the stockman actually raises . . . the cattle are only the method by which the commodity is merchandised. Cattle (or other grazing animals) should always be given second priority to the forage. In actual practice, the reverse is all too often the case. As a result, a large percentage of the available rangeland has been overgrazed, and is therefore producing less than its potential. Depending upon the overall climate and available moisture, this damage takes years, or even decades to repair.

EFFECTS OF OVERGRAZING

Overgrazing eventually kills the individual plants by reducing the leaf area to a point that it cannot function adequately. Plants receive their elemental nutrients and water from the soil (roots), and transfer it to the leaves where photosynthesis takes CO_2

from the air and forms the carbohydrates that the plant uses for energy. Excess carbohydrate is then stored in the roots or the crown of the plant as an energy reserve (Figure 2-2).

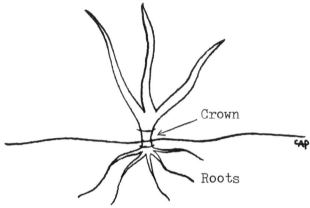

Figure 2-2. CARBOHYDRATE STORAGE AREAS IN GRASS PLANTS.

If the leaves of a plant are grazed, then a reduced amount of leaf area will be left to carry out photosynthesis. Blades of grass, of course, are the leaves of the plant. As a general rule, if more than 50% of the natural length of grass is removed, the plant will not be able to produce enough carbohydrate to meet its daily needs (Ragsdale, 1971). The plant will then call on the reserves in the crown and roots to make up the deficit. New leaf growth is initiated, and if undisturbed, the plant will again have enough leaf surface area to photosynthesize more than enough carbohydrate to meet daily needs. After a period of time, the plant will produce a surplus of carbohydrate so that excess may again be stored in the crown and roots.

But if the new leaf growth is continually removed (as in heavy grazing), then the plant will be forced to rely on its reserves for the carbohydrate it requires. Eventually the plant reserves will be exhausted, and the plant will die.

PLANT SUCCESSION
Whenever a plant dies out, a new plant species will move in to take its place. The new species doesn't actually come from another area. Usually it is a plant that is already present, but has been supressed by the more dominant plant. That is, "survival of the fittest" is just as applicable to the plant world as it is the animal world. For a given set of environmental conditions there will be one plant which will be the most adapted. That plant will

therefore be able to out-compete all other plants for what nutrients are available in that particular environmental setting. It is important to realize that sunlight is a nutrient, and therefore tall growing plants will literally "over shadow" shorter plants and therefore have an advantage.

As a general rule grasses are able to out-compete most other plant types. The biggest exception to that rule are forest habitats where the soil may be too acid, top soil too thin or sparsely distributed (frequent rock outcroppings), etc. Once established, evergreen or other trees overshadow the soil and further plant growth (including trees) is inhibited. But under most environmental conditions grasses will predominate. Normally the tallest grass that the environment will support will be dominant. Again, shading becomes a factor.

If that grass is overgrazed and killed out, a shorter grass will usually take its place. If that grass is overgrazed, an even shorter grass usually takes its place. (Figure 2-3) If grazing pressure continues to be severe, eventually a woody or otherwise unpalatable species will become the dominant plant.

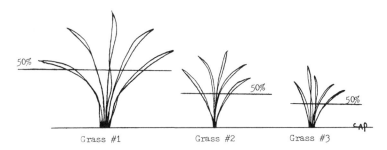

Figure 2-3. PLANT SUCCESSION DUE TO OVERGRAZING.

The classic examples of this are sagebrush and mesquite. They are both relatively unpalatable and have therefore become very prevalent in overgrazed areas.

According to records and accounts by surviving pioneers, the desert areas of the southwestern and northwestern United States were once grasslands interlaced with sparse populations of sage and mesquite. There was very little change until the advent of barbed wire and windmills. Before, livestock had to be moved fairly rapidly from water hole to water hole, and so only the areas around the water became overgrazed. But when it became possible to confine animals continuously, overgrazing took its toll. As the grass was grazed out, the unpalatable plant species became the dominant plants. Essentially, this is how the sagebrush deserts of the Northwest Basin and the mesquite deserts of the Southwest came to be.

HOW OVERGRAZING USUALLY OCCURS

The overstocking of ranches is usually not intentional. Rather, what typically happens is the rancher stocks the number of animals that his ranch will support during a good year. When rainfall is less than adequate, most ranchers are hesitant to cut back on their herd. The primary reason being that during drouth years the cattle market is usually depressed. This is because other ranchers don't want to buy range cows as they don't have enough grass either. Thus, the only buyers for their cows are packers, or feedlots. Obviously, packers and feedlots are not willing to pay for the genetic value of a cow, only the meat value. The situation creates a buyer's market.

Most ranchers will keep hoping that rain will come soon. They know that with rain will come higher cattle prices, and so they hold on. Typically they will not sell any cattle (cows), until they have no available forage left. When the cattle get to the point of starvation, most ranchers will either sell, or begin providing extra feed (hay, etc.). This further stresses the range simply by keeping the cattle on it longer. The amount of feed provided is usually minimal, and so the cattle will continue to utilize every bit of available forage.

The real damage, however, will occur when the rains finally do come. When moisture is once again made available to the range, the plants will begin to grow again. Since the plants have been dormant, nutrients for new growth must come from reserves within the plant, the crown (if any material is left above the ground), and/or the roots.

The situation is that lush regrowth is suddenly made available to a cow herd that is half starved and has not had one sprig of green grass in months, or even years. Cattle, cows in particular, can eat enormous quantities of feed for short periods when they have been deprived previously. A herd of cows under these circumstances can eat 50 to 80 lb of range forage for several days or even weeks. By taking and retaking the regrowth, the plant reserves will soon be exhausted, the plant will die, and the succession process explained in the previous section will occur.

RESTORATION OF OVERGRAZED RANGELANDS

The terrible thing about overgrazing is that some rangelands can never be returned to normal . . . because overgrazing not only increases the incidence of undesirable plants, it also often initiates erosion. High quality topsoil is gone. Topsoil, of course, takes eons of time to form.

Where erosion has not been a factor, plant succession will work in reverse. Under proper grazing management, longer and longer grasses will return. The question and problem is one of time.

The amount of available moisture is the determining factor in dictating the time required to restore an overgrazed range. The greater the available moisture, the quicker the range will be restored; i.e. the faster plant succession can work. The sad part about it is that the majority of areas that have been overgrazed have an arid climate.

Total precipitation, of course, greatly influences the amount of available moisture, but the timing of precipitation is also vitally important. For moisture to be truly effective, it must come or be available during the growing season. In the Southwestern United States the majority of the precipitation comes as rainfall during early summer and fall. In the Northwest most of the precipitation comes as snow during the winter, and so there is far less moisture actually available for plant growth. This, coupled with the short growing season, greatly hinders range improvement. Specifically, it takes several times as long to restore the sagebrush deserts of the Great Basin than the mesquite covered deserts of the Southwest.

Range seeding as an alternative to natural restoration has been practiced with both success and failure. While there are a myriad of factors involved, available moisture is again the main determining factor. In relatively humid areas the main obstacle is primarily a mechanical one of removing the overstory brush and seeding the ground. If the area has adequate rainfall, the success of the reseeding is more of an economic question than a physical one; i.e. with proper technical direction, brush can be removed and grasses made to grow.

But as mentioned earlier, the real problem areas have arid climates. In the Southwest, it was said that most of the precipitation falls during the growing season, which makes rangelands easier to restore through natural restoration (than the Northwest). However for the purposes of reseeding there is another factor that comes into play . . . and that is the unreliability of rainfall. The Southwest is notorious for long dry spells, often ended with flash floods.

Reseeding Southwestern desert country is very risky business, as one cannot depend on the rainfall. Normally some type of brush removal is required, in order to form a seed bed for the grass. If a gentle rain comes soon after, then the grass will germinate and the seeding will be a success. But if the rain does not come; more than likely hot, dry winds will. There is a good chance then, that the seed will be blown away and the area will be eroded. Then too, a torrential rain is always a possibility, which would also erode the land and remove the seed. Seed can be broadcast without disturbing the soil, but the possibility of wind or rain carrying it away is even greater. There is no increased danger of erosion with broadcasting, but at the same time the amount of seed that will germinate and become established (under favorable conditions) will be greatly reduced. This is not to say that there have not been cases of success in reseeding arid areas in the Southwest, because there have. It should be pointed out, however, that seeding done in this area is usually done with either public funds or with tax oriented risk capital. The risk is generally too'great for the bonafide rancher seeking return only in the form of increased grazing.

Greater successes with reseeding have been recorded in the Northwestern desert areas. This has essentially been due to the development of two winter hardy range grass species, crested wheatgrass, and Russian rye. As mentioned earlier, most of the precipitation in the Northwest comes as snow in the winter, and so the bulk of the moisture is not available to native grasses that grow during the temperate months. But with the introduction of these winter growing grasses, this moisture can be utilized. Unlike seeding in the Southwest, there is not near the erosion danger. Seeded in the fall, precipitation comes in the form of snow, and therefore there is no danger of water erosion. High winds may blow in the winter but usually they accompany snow that will cover the ground and tend to protect it. Springtime can bring hot dry winds, but by that time the grass should be established. Though not as risky as seeding in the Southwest, brush removal and seeding in the Northwest is still very expensive. Extensive reseeding on private Northwestern ranches was practiced in the 50's and 60's, but decreased considerably after that. It is the author's opinion that this is due to the Tax Reform Act of 1969 which specifies that land development expenses must be capitalized. Prior to 1969 land development expenses were deductible in the year of origin (see page 172). Again it is difficult to justify range seeding on the basis of increased grazing only.

The point of this whole discussion has actually been to point out the very serious damage that overgrazing creates. In the majority of the western ranching areas with arid climates, the damage may be irreparable, for all practical purposes. Erosion may make restoration physically impossible; natural restoration through reduced stocking may take more years than the rancher can afford; and range reseeding may not be justifiable economically.

RANGELAND GRAZING SYSTEMS

The purpose of this section will be to familiarize the reader with basic concepts concerning grazing systems. Only generalities will be covered as specific details concerning plant taxonomy, botany, ecology, and soil science are beyond the scope of this text. Because of the myriads of interacting details, this section has been prepared to simply help the reader recognize the need for planned grazing systems. To actually develop a system, it is advised that a suitable range technician be consulted (range extension agent, SCS specialist, private consultant, etc.).

CONTINUOUS GRAZING

Continuous grazing implies livestock left in an enclosure year round. It takes the least amount of management effort and capital (for fencing), and so it is the most popular grazing system. However all too often it is not practiced as a system. Rather, the cattle are simply "turned out".

The biggest problem associated with continuous grazing is what is known as spot grazing. As the name implies, this means that grazing animals will graze individual spots within an enclosure while leaving other areas ungrazed. The spots that are grazed and regrazed will obviously have much more immature growth on them than surrounding plants. That is, after the plants are grazed, regrowth will occur. Regrowth, of course, is very green and lush, and is much more palatable than more mature growth.

When this new growth appears, the animals will regraze that spot. In areas grazed continuously for a long time, there will be a number of areas or spots that the animals will continually return to. It is a vicious cycle. If stocking is light, the main deleterious effect will simply be that only a portion of the range will be utilized. The vegetation in the spots will be grazed, and the vegetation outside of the spots will become mature and will not be utilized. Tall, mature forage is much higher in fiber, lower in protein and digestible carbohydrates, and much less palatable than lush green, immature forage.

In areas with considerable rainfall the grazing spots will be fairly evident (patches of green surrounded by yellow and brown). In more arid areas with scattered vegetation, the spots may not be as evident. They may not appear as spots at all. Instead, one may simply find an intermixture of tall yellow grass with short green grass. Often the tall grass will be clump or bunch grass.

At the very least, continuous grazing will result in incomplete utilization of the range. In most cases the spot grazing will result in the vegetation within the spots being grazed out. A good range taxonomist can usually tell if an area had been grazed continuously for very long as the more palatable plants native to that area will be missing. The most palatable plants are usually the most nutritious, and so, as a general rule, it can be said that continuous grazing will not only reduce the utilization of the range, but if carried over a long period of time, will actually reduce the productivity of the land.

SEASONAL GRAZING

Seasonal grazing typically implies the grazing of pastures which can only be used during certain seasons of the year. Most commonly, it refers to mountain pastures which can only be used during the summer (due to ice and snow at other times).

GRAZING SYSTEMS UTILIZING MULTIPLE PASTURES FOR THE SAME HERD OF CATTLE

As the name implies, multiple pasture systems means splitting the range up into smaller plots and grazing them in some sort of alternate pattern. At some point during the course of the grazing program, at least one of the pastures will be set aside and not grazed.

TERMS USED IN MULTIPLE PASTURE SYSTEMS

The terms used to describe multiple pasture systems can be confusing as they often refer to systems which are quite similar. Essentially there are two basic types; (1.) rotational systems, and (2.) deferred systems. Most other systems are modifications or combinations of these two basic types. The following is a description of the most commonly used systems.

Rotational Grazing—Rotational grazing commonly refers to grazing the subdivided pastures alternately. The basic idea being that by concentrating a large number of animals in a smaller area (subdivided pasture), a more even utilization of forage is obtained. Rather than utilizing only the most lush or otherwise palatable forage, the animals are forced to consume a larger amount of the less palatable and more mature vegetation. There is less trampling since the animals do not have to move around as much to graze. Likewise, areas around water do not get damaged as much.

Deferred Grazing—Deferred grazing usually refers to leaving a given pasture ungrazed for a specific season or time period. The actual time period set aside for deferment will be determined by the type of vegetation contained in the pasture. For instance, if the most important plants in a pasture are propagated by seeds, the pasture would be deferred until after the seed set.

Probably the most basic difference between deferred grazing and rotational grazing is that deferred grazing puts emphasis upon individual plant species, whereas rotational grazing does not. Pastures are deferred to benefit individuall plants, whereas pastures are rotated to benefit all the forage.

Deferred-Rotational Grazing—Obviously this is a combination of both the deferred and the rotation systems. At some time various pastures may be deferred, and then later rotated. That is, pastures are put together in blocks or range subunits. The pastures in one block may be entirely deferred, whereas pastures in other blocks are grazed rotationally.

Rest-Rotation Grazing—Rest-rotation grazing is a relatively complicated system used by the Forest Service. Rest is defined to be the deferment of pasture for a full year. Each year one pasture is usually rested, while the others are grazed and rotationally deferred.

High Intensity, Low Frequency Grazing—This is an intensive plan of rotational grazing developed by Dr. Merrill at the Texas Experiment Station. With this system the range is split into several small pastures which are grazed individually by a large number of animals for a short period; sometimes for a period as short as two weeks.

This system is well adapted to brushy areas, as it forces the animals to utilize all the available forage, including browse. This takes pressure off the more palatable plants, but may stress some of the forbs.

CONCLUSIONS

The purpose of this section has been to point out the importance of range management. The author sincerely hopes that the reader will recognize and remember that range forage is the commodity that the rancher actually produces, and that livestock are merely the method of marketing the forage. Clearly, ranching decisions should always give priority to the vegetation.

Because of the many self-proclaimed environmental experts

in recent years, ranchers as a whole have become disdainful of the words ecology and environmentalist. Proper range management is not merely a matter of ecology or environmentalism, it is a matter of economics and survival. In essence, making decisions in favor of livestock (over the vegetation), is short-sighted, and can be disastrous in the long run. As mentioned earlier, 50% utilization of range grass is used as a common rule of thumb* (Ragsdale, 1971). Put in the words of a knowledgeable rancher, "When it comes to grass . . . take half, and leave half - and you'll have a bigger half next year.".

TYPE OF OPERATION - STOCKER VS. MOTHER COW

Ranching consists of either stocker or mother cow operations, and sometimes a combination of the two. The most common type of operation is the mother cow. This simply means that the rancher runs a breeding herd of cows, and sells calves produced from the herd. Stocker operations consist of taking weaned calves and running them on pasture to be sold as yearlings. Starting weights for the calves will normally run from 300 to 500 lb, and ending weights will usually run from about 550 to 800 lb. Steers are most commonly run, and hence the term stocker-steer is often used as a term meant to be synonymous with "stocker" operations.

What type of enterprise a ranch should become involved in depends upon a number of factors. The most basic factor is the quality of the range itself. In order to have a successful stocker operation, the ranch must be capable of producing gains of at least 1 lb/hd/day. Breeding cows need only to be maintained (do not need to gain weight), and therefore do not require nearly as good a quality of range. Since the majority of the rangeland in the western U.S. is somewhat desolate, it is easy to see why cow-calf operations are the most common type of ranch enterprise. There are, of course, periods of time during the year when nearly all ranches can produce weight gains on their animals (spring and/or fall), and this is when the rancher usually plans to have his cows calve. Stocker operations must provide for weight gains through a much longer time period.

For ranches that are physically capable of either stocker or cow-calf operations, the major consideration is economics. There are a number of economic differences, but by far the most crucial

---

*Actual optimum utilization rates can vary greatly, depending upon the situation.

difference is that in stocker operations the entire investment is bought and sold in a one year period or less. This exposes the stocker operation to much more market risk (see also p.67).

SPRING VS. FALL CALVING

Feed requirements for cows essentially double as soon as they begin lactating. Grasses and forages typically have their highest nutritive value in the spring, and so by no small coincidence, most wild animals bear their young in the spring. In addition the milder spring weather is also more conducive to offspring survival.

For these reasons, cow herds have traditionally been bred to calve in the spring. The calves are kept on the cow through the summer and when the forages begin to decline in nutritive value in the fall, the calves are weaned. For the physical process of raising cattle, this plan works the best for most of the western ranching areas. Marketing problems have occurred, however, when all the calves come off in the fall, thereby glutting the market.

To avoid what was typically a low calf market in the fall, many producers in the southwestern part of the U.S. went to fall calving. This allowed them to market their calves in the spring or summer, and thereby helped to even out the calf market. In actual practice, fall calving works just about as well as spring calving, and in some areas works better. In areas such as South Texas, and the lowlands of New Mexico, Arizona, and California, fall calving lets the calves avoid the flies and 100°F temperatures of the summer. Winter grasses aren't any poorer in quality than the typically burned up summer grasses, and so fall calving can sometimes be more advantageous.

In recent years, the emergence of the Southeast as a major cow-calf producing area has reduced seasonal marketing problems. Both fall and spring calving are practiced, as well as year round calving in many of the small operations (bulls are left with the cows year 'round). This, plus widespread early weaning (weaning after 3 or 4 months instead of the traditional 7), has made calves available nearly all year, and thus greatly reduced the seasonal marketing problems that existed earlier. There are still more calves marketed during the fall, but because of the tax advantages to cattle feeding, the market slump is not nearly as severe; i.e. because of increased end of the year buying (page 164). Then too, the advent of fall and winter grazing of winter wheat in the Great Plains has also increased the demand for fall calves.

THE YEARLY PRODUCTION CYCLE (For Spring Calving)

One might say the production cycle starts at calving, or at breeding time. For the purpose of this section, the yearly production cycle will be said to start at calving time. Calving, of course, may take place during the spring or the fall, depending upon whatever production plan the ranch is following. Spring calving is still the most popular plan, so it will be a spring calving program that is described here. A fall calving program would be similar, only the timing would be 6 months later.

Calving will begin near the end of February and last until the end of May. Occasionally some late calves may be born in early June, but most operations try to avoid such late calving. First calf heifers will have been brought up to some type of small enclosed pasture or field in order that they may be observed more easily. Moreover, they should be observed at least once every 8 hr. With first calf heifers a certain amount of calving difficulty can be expected to occur, and so management should be ready to cope with it when it occurs (see page 271). The cow herd may or may not be brought in during the calving season.

Breeding season will normally run from the 1st of June until the middle of August. Cows cycle on an average of 21 days, so most breeding seasons are designed to allow for 3 heat cycles. As mentioned on page 277, plane of nutrition greatly influences the onset of estrus. Since lactation has priority over estrus for nutrients (calves born in the spring are nursing their dams - which now must be rebred), it is of vital importance that the cow herd receive adequate nutrients. On rangeland, since summer's heat tends to dry up and make forage less nutritious, a protein supplement is often put out (see page 29).

Since lactating cows have nutrient requirements much higher than dry cows (8 lb TDN vs. 12-16 lb TDN), some operations wean their calves either just before or during the breeding season; i.e. when the calves are 3 to 4 months old. This greatly increases the amount of energy available for estrus, and therefore results in higher calf crop percentages (see page 28). If early weaning is practiced, it would take place anytime from the 1st of June up until about mid-July. Actual dates would depend on the considerations discussed on page 29.

Sometime during the late spring or early summer, most ranches will gather their cattle for the purpose of branding and vaccinating the calves. On some ranches, branding may extend on into mid-summer. At the same time, the cow herd is often dipped, sprayed, or treated with a pour-on grubacide.

Most ranches in the northern ranching areas will have some

Figure 2-4. YEARLY PRODUCTION CYCLES.

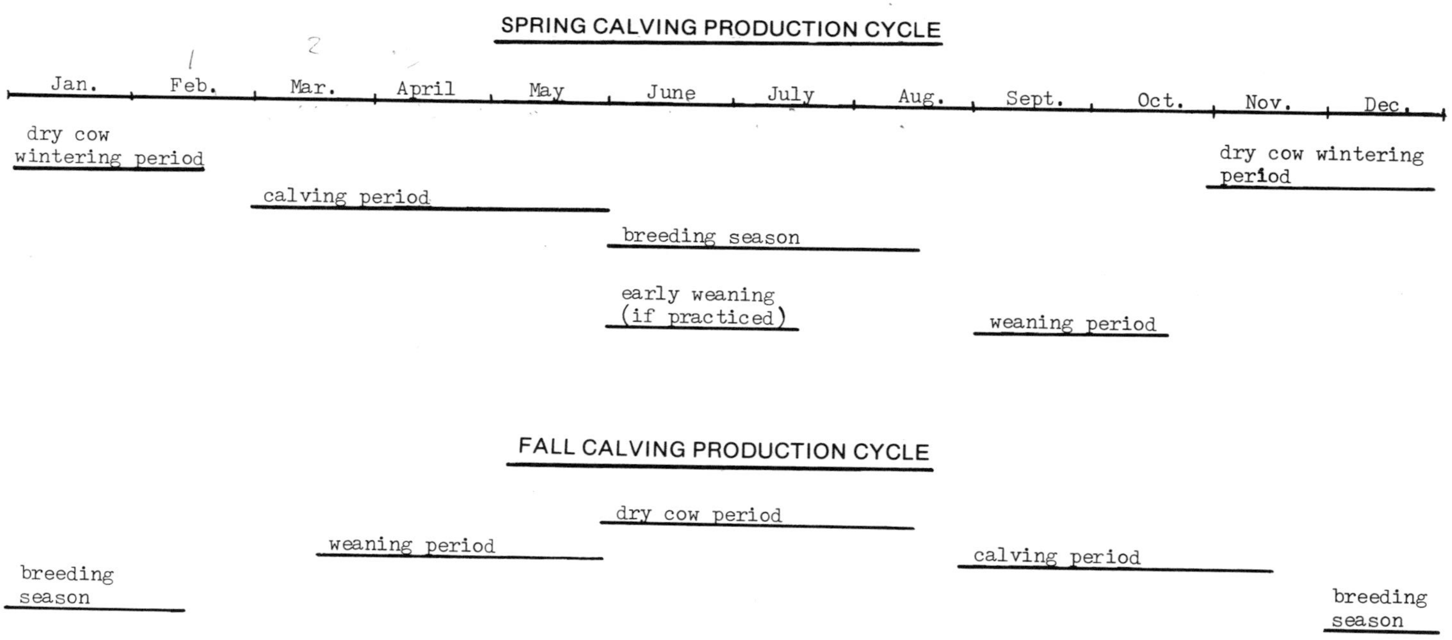

SPRING CALVING PRODUCTION CYCLE

Jan.　Feb.　Mar.　April　May　June　July　Aug.　Sept.　Oct.　Nov.　Dec.

dry cow wintering period

calving period

breeding season

early weaning (if practiced)

weaning period

dry cow wintering period

FALL CALVING PRODUCTION CYCLE

breeding season

weaning period

dry cow period

calving period

breeding season

meadows or fields which are used for producing hay and sometimes silage for wintering the cow herd. Late spring and early summer marks the haying season, and thus creates a goodly demand for the ranch labor.

Weaning will generally take place sometime between August and November. Normally, most calves are weaned between early September and mid-October. At the same time the calves are weaned, most progressive operations will also pregnancy test the cow herd before turning them out again.

After weaning, the cow herd is usually left to graze until the onset of winter. In areas where winters are severe, the cows are brought in and kept in lowland meadows, valleys, or fields, and fed the hay and/or silage put up during the summer. In areas where winters are not severe, the cow herd is usually left out on the range, and supplemented as required.

Through the winter the only major labor requirement is feeding and/or supplementation, depending upon the area. For that reason, repairs to ranch machinery, fences, etc. are normally conducted during the winter.

Winter's end brings the onset of calving, which begins another production cycle.

SELECTION OF BREEDS FOR COW-CALF RANCHING

For specific information, the reader is referred to Chap. 15, which discusses individual breeds in detail. For the purposes of this chapter, the author would simply like to point out that the selection of a breed or breeds should be based on hard facts. In far too many cases, the selection of a breed has been based on personal preference or tradition.

The most common mistake in commercial herds is the failure to take advantage of crossbreeding. Explained in detail beginning on page 240, crossbreeding is a principle known to increase percent calf crop and weaning weight. McPeak (1980) summarized data from crossbreeding studies with Angus, Hereford, and Shorthorn cattle (Table 2-1).

Table 2-1. CROSSBREEDING DATA.

Trait	Straightbred	Straightbred cow crossbred calf	Crossbred cow crossbred calf
Calf crop weaned	80%	84.1%	88.8%
Weaning wt	430 lb	450 lb	474 lb

For commercial herds,* in most situations the question of breed selection is which breeds to include in a crossbreeding system.

For purebred operations (production of seed stock), the individual breed selected will depend primarily on market demand and adaptability to the local environment. In the past, breed preferences and subsequent market demand have fluctuated in the form of fads. This has unfortunately placed as much or more emphasis on the promotion capabilities of a breeder, as it has his ability to produce cattle of sound genetic quality. The first fad was the "blocky, compact" cattle of the 1930's and 40's, and the most recent fad was the so called "exotic breeds" of continental Europe. Promotion and a certain amount of "fadism" will always be a part of the business. However, it is the author's opinion that in the future there will be a movement toward a growthy type of cattle that produce a lean carcass . . . and this movement will not be a fad, so much as it will be a logical, methodical trend. A trend brought on by the changing consumer demand for hamburger and lean cuts. The only thing that is currently holding the industry back are the marbling standards set for the choice quality grade. At the present time there are groups within the industry who are attempting to bring pressure upon the USDA to change the quality standards. If they are successful, the move toward lean, growthy cattle will come quickly. If not, it will come more slowly, but it will come.

It is therefore the author's opinion that purebred breeders should design their breeding programs along these lines. This does not mean moving to or away from individual breeds so much as it means a change in selection procedures within those breeds. The commercial market will demand cattle that work well in crossbreeding programs, and since crossbreeding needs cattle of diverse genetic backgrounds, there will be a need for many different breeds. But when the commercial breeder chooses a breed, it will be on the basis of its genetic worth . . . what it can do to increase the productivity of his herd. No longer will cattle be selected on the basis of whether they do or do not have spots, solid or particular coat colors, etc. Cattle will be selected on the basis of productivity and this is what the purebred breeder should select for.

---

*Ranches that are not in the business of selling breeding stock; i.e. sell animals only for their meat value.

BREEDING PROGRAMS FOR COW-CALF RANCHING

The three types of breeding; pasture breeding, hand mating, and artificial insemination are discussed at length in Chapter 14. For the purposes of this chapter, the author deems it appropriate to make two suggestions; that each and every rancher should: (1.) give serious consideration to artificial insemination (AI), (2.) keep accurate breeding records. For many operations AI may not be feasible, but to determine whether it is, or is not feasible, the rancher should look at the actual costs and returns (see page 238). Breeding records are a necessity for A.I. but regardless of the type of breeding program utilized, records should be kept. Oftentimes much of the increased return from A.I. is due to record keeping alone, because the keeping of records makes ranchers much more conscious of their overall management.

CREEP FEEDING

Creep feeding is defined as feeding a concentrate to calves while they are still nursing their mothers. It is usually accomplished by using a self feeder surrounded by a cage of bars that are set far enough apart to allow access by calves, but not cows.

Creep feeding will obviously increase calf weaning weight. Whether the practice is economical or not depends on a number of factors, some of which may or may not be capable of being determined.

The simplistic way of determining the economics of creep feeding has been to figure how much feed the calves ate or will eat, and compare that with a projected increased weaning weight at a constant price. For example, if calves eat 2.0 lb of creep feed a day costing 6¢/lb for 120 days; weigh 440 lb instead of the usual 400 lb at weaning; and sell for 60¢; the first inclination is usually to figure the cost of gain and return on that basis. It would seem then that the return from creep feeding was $9.60/hd.

```
120 days
x 2 lb   consumption
240 lb   total consumption
x $.06   cost per lb
$14.40   total feed cost
```

Non-creep Calves   Creep Calves

400 lb weaning wt   440 lb weaning wt

x .60 sale price   x .60 sale price

$240   per hd sale price $264   per hd sale price

$264   creep calf sale price

-240   non-creep calf sale price

$24   difference

$14.40 feed cost

$9.60   return from creep feeding

The fallacy in this reasoning that the cattleman will quickly see is that the sale price on a per pound basis will not be the same for 400 lb calves as it is for 440 lb calves. Different weight cattle bring different prices. Lighter cattle bring a higher price than heavier cattle; particularly when the weight is from extra fat, as it would be in the case of creep fed calves.

Using this example, calves of the same frame with 40 lb less flesh, would ordinarily bring a 3 or 4¢ premium. At a 3¢ premium creep feeding would appear to be a losing proposition.

Creeped calves   Non-creep Calves

440 lb   400 lb

x .60   x .63

$264 $252

$264

-252

$12

$14.40   feed cost

-($2.40) advantage to creep feeding

What this example doesn't take into consideration is how much less forage the creeped calves ate. This figure is difficult if not impossible to obtain.

The intent of this section has been to point out how perplexing the decision to creep feed or not can be. In a review of real-life experiments, Preston and Willis (1970) found no continuity among creep feeding trials. Some showed it was profitable, some did not.

Probably the only situation where creep feeding would normally always be of benefit would be in an early weaning program. If calves are used to eating concentrates out of a feeder, weaning is usually less traumatic.

WEANING

Figure 2-5. The idea behind weaning is to get calves to forget about bawling for mamma, and begin eating a mixed ration or other feed as soon as possible. Otherwise, a nutritional stress will occur which can aggravate sickness and death loss due to shipping fever.

Weaning is simply defined as taking the calf away from the cow; the calves are separated from their dams to prevent nursing.

This places emotional stress on the calves, and can result in considerable sickness and death loss unless handled properly.

When separated, cows and calves should be placed far enough apart so they are out of earshot of one another. On a clear, calm day, this can be further than one might expect. Cows' and calves' bawling can be heard over a suprisingly long distance. As long as the cows can hear their calves bawl, they will bawl back, and the longer the calves can hear their mother, the longer and more difficult the weaning period will be.

The whole point of weaning is to get the calves to forget about mamma, and start eating a mixed ration, or whatever other feed is provided. The longer the calves bawl, the less they will eat. The less they eat, the weaker they will become, and the greater the sickness and death losses will be. (See also, backgrounding, page 53.)

EARLY WEANING

When early weaning is practiced, it is usually done in the hopes of obtaining a higher rebreeding conception rate, although it is also practiced in some instances to increase carrying capacity of the ranch or farm. Moreover, it is usually adverse pasture conditions that indicate the use of early weaning.

The breeding season starts 2-3 months after calving. Under typical cow-calf management, calves are weaned at about 7 months. With most cow-calf operations then, calves are suckling cows throughout the breeding season.

Having calves suckling cows during the breeding season can have detrimental effects on rebreeding. To begin with, Wiltbank (1958), has shown that suckling tends to inhibit the ovarian cycle. Probably more important on most ranches is the fact that lactation greatly increases the nutrient requirements (8 lb TDN for a dry cow vs. 12-16 lb TDN for lactating cows). Since lactation has priority for nutrients over estrus (page 278), nutrition becomes of grave importance.

Adding emphasis to the problem of nutrition is the fact that breeding seasons typically occur during the poorer growing months for forage. Spring calving programs would have a summer breeding season, and fall calving programs a winter breeding season.

To circumvent these problems, some operations make it a practice to wean some or all of their calves just before or during breeding season. Not only does this tend to increase the conception rate but also increases the carrying capacity of the ranch or farm. Since dry cows require 40 to 60% less feed than lactating cows, a larger number of cows may be carried.

The actual time of weaning will usually depend on pasture conditions. Weaning before the breeding season will catch calves from 1 to 3½ months old. Weaning calves under 2 months of age is a difficult task to accomplish without excessive death losses. If pasture conditions permit, it is often a better plan to wait and wean a month or so after the breeding season begins, or at least delay the weaning on the younger calves. In spring calving programs, this would coincide with the month of July, and in many places forage quality remains relatively good up until July. Even by delaying weaning until all calves are 3 to 4 months old, however, substantial health problems are created, as the stress of weaning is still much greater than with normal weaning at 6 to 8 months. Rather than try to cope with the health problem, many operations (particularly in the Southeast) send their calves directly to market (auction barns, livestock trader pens, etc.). Ideally this is not the way to properly handle young calves, but it is done, and this is a primary reason why shipping fever losses are so high.

In actual practice there are very few operations that early wean their entire herd. More common (among progressive operations), is the early weaning of only the first calf heifers' calves, as in order to rebreed they not only have to overcome the nutritional demands of lactation, but also their growth requirements. The results of a Texas experiment involving the early weaning of two year old heifers are shown in Table 2 - 2 (McCartor, 1972).

Table 2-2. REBREEDING PERFORMANCE OF TWO YEAR OLD HEIFERS EARLY WEANED, OR LEFT W/SUCKLING CALVES.

% Conception	Suckled Heifers	Early Weaned
at first estrus	36.7%	67.0%
at end of breeding season	55.2	84.8

THE NEED FOR SUPPLEMENTATION OF RANGE COWS

Supplemental feed usually accounts for the largest single cash outlay in ranching. For that reason the author deemed it necessary to discuss the different aspects in detail.

When the grass and other forage is green and actively growing, it is relatively high in protein and energy (soluble carbohydrates).

Green, lush grass can be as high as 25-30% crude protein, although 12-18% would be more common. When growing conditions deteriorate (onset of winter or summer, heat, drouth, etc.), and the grass turns from green to yellow or brown, the nutritive value drops off very quickly. Protein contents can go as low as 4%, and TDN can go from 65-70%, to 45 and 50%.

When conditions such as these occur, beef cows must be supplemented if they are to remain on the pasture. Supplementation is defined as providing a small amount of a concentrate in order to allow the cattle to more fully utilize the range forage that is available. This is opposed to range feeding which is defined as providing a large amount of the animal's total intake from a feed source other than range forage.

The distinction between supplementation and feeding may seem to be a matter of semantics . . . it is not! One of the most crucial questions to any ranch is, "Are we engaged in a feeding program, or a supplementation program?". The basic enterprise in ranching is to produce range forage. Indeed, the range forage is, or at least should be, the cheapest feed available to the rancher. Under nearly all historic economic conditions, feeding harvested feed to beef cows has been a losing proposition. In addition, whenever cattle must routinely be "fed" on rangeland, the ranch is obviously overstocked. At some point then, overgrazing will almost always occur.

The strategy should be to supplement as little concentrate feed as possible. To understand what and how to supplement, one must understand why supplementation is so important.

When the protein and energy values of grasses fall off, the fiber content goes up. Fiber is difficult to digest, which reduces consumption considerably.

Essentially, the problem is this: A dry, pregnant cow needs about 8.0 lb of TDN to maintain her bodyweight and produce a healthy calf. Dry, mature range grass will have a TDN which may average about 47%. Without supplementation, a cow will eat about 1.75% her bodyweight in this type of material. A 900 lb cow will therefore eat about 15.75 lb (900 x .0175) of that kind of grass. As explained on page 195, the reason for the decreased intake is the build up of fiber in the rumen; i.e. the fiber is difficult to digest, it builds up in the rumen and the cow feels "full". With an intake of 15.75 lb and a TDN of 47%, the cow's TDN intake is only 7.4 lb (.47 x 15.75). Since she needs 8.0 lb TDN, but only receives 7.4 lb TDN, the cow will obviously lose weight.

More important than the weight loss, are the adverse effects upon reproductive efficiency. These are caused by what are known as priorities for nutrients. This means that certain physiological functions are given priorities for available nutrients. The priorities that pertain to reproductive functions are (in order of priority):

[
      I. Lactation
      II. Body Maintenance
      III. Fetal Growth, or
      III. Estrous Cycle

Lactation is given priority over all other body functions. The cow will deplete her own body stores of energy, protein, and minerals in order to meet her genetic capability to give milk. This is why the cows that produce the heaviest calves at weaning will often be in the poorest condition.

After lactation, body maintenance is given priority. Body maintenance is given preference over estrus and fetal growth, although the division is not as clear-cut as with lactation. In borderline cases of underfeeding, the fetus will continue to grow, and a small percentage of the cows in the herd will fail to come into heat. If malnutrition is substantial the fetus will be born small and weak, or in extreme cases, may be resorbed. (Underfeeding is the most critical during the last three months of pregnancy.) Likewise, a disastrously large percentage of cows may fail to breed if serious malnutrition occurs during the breeding season (especially if under nutrition also occurred during pregnancy).

In terms of survival of the species, these phenomena seem easy to rationalize. In the case of lactation and body maintenance being given priority over fetal growth and estrus, it seems logical to conclude that nature has in effect given priority to the newborn animal, saying "Look, we've got a calf on the ground, let's take care of it, and then if there is anything left over, we'll see about making another one". In the case of body maintenance given priority over fetal growth, it seems obvious that a weak or dying cow (from starvation), cannot give birth to a live calf.

WHAT TO SUPPLEMENT

Since range cows cannot meet their TDN requirements from dry, mature, range grass, it should be apparent that some type of supplementation is required. The question is, "What kind of supplement?".

Since it is energy that controls estrus (see page 277), many stockmen have reasoned that it is energy (grain) that they should supplement with. While feeding grain will certainly help cows to cycle, it is relatively inefficient and usually uneconomical. In the example given, .6 additional pounds of TDN are needed (cow needs 8.0 lb and is getting 7.4 lb). Theoretically, less than 1 lb of grain would be needed to rectify the situation. In actual practice, it would take from 3 to 6 lb of grain to keep a range cow in good shape.

There are two reasons for this. The primary reason is easily understood. Grain is much more palatable than dried up grass, and so range cattle will reduce their consumption of grass. Moreover, if they know that they get fed grain at a particular time, they will quit grazing early, and go to the area where they are usually fed. If they are fed with a self feeder utilizing some sort of intake inhibitor (salt, gypsum, etc.), they will hang around the feeder at least 2 or 3 hr a day.

The second reason is more complex. Rumen microorganisms which do the actual digesting (in the rumen), develop specific populations for certain feeds. In general, there are two types, cellulose users (cellulolytic), and starch users (amylolytic). If cattle are consuming feeds high in fiber, a population of cellulose using bacteria will develop. If cattle are fed a grain ration, a population of starch users will develop. If cattle are fed both fiber and starch intermittently the rumen population will be in constant turmoil, and a good population of neither kind of microorganism will develop. The end result will be inefficient utilization of both feeds.

The feeding of small amounts of protein, however, usually results in an increased forage intake. As explained on page 195, low quality roughage is quite low in digestible protein . . . so low, that the rumen microbial population often cannot be maintained on it. Supplementation with a small amount of protein allows the microorganisms to reproduce, and thereby increases the total number of microorganisms available for digestion. With more organisms working on the fiber present in the rumen, it is broken down and digested faster. This makes the cow feel less full, and therefore she is able to eat more forage.

With proper protein supplementation, intake of low quality roughage will increase from 10 to as much as 50%. Going back to the example of the cow that eats only 15.75 lb of dry grass, the advantage of only a 10% increase in intake can readily be seen; i.e. if the cow now eats 17.3 lb of a 47% TDN grass, she will consume 8.1 lb TDN (instead of 7.4), and will therefore meet her requirement (8.0 lb of TDN).

Along with helping to meet energy requirements through increased forage intake, protein also quite obviously helps to eliminate the protein deficiency created by grazing low quality forage. Protein deficiencies during gestation (page 277) can result in decreased fetal growth.

Table 2-3a and 2-3b represent the results of feeding a protein supplement to heifers fed poor quality hay.

Table 2-3a. EFFECT OF SUPPLEMENTATION ON PERFORMANCE OF WEANER HEIFER CALVES FED PRAIRIE HAY.*

	None	Daily Supplement 1.25 lb CSM	1.25 lb Corn
No. heifers	10	10	10
Avg. gain, lb	-115	35	-92
Avg. hay intake, lb/day	8.9	11.9	8.9

Table 2-3b. EFFECT OF PROTEIN SUPPLEMENTATION ON PER – FORMANCE OF PREGNANT HEIFERS WINTERED ON PRAIRIE HAY.*

	Cottonseed Meal	
	None to calving, 1.0 lb/day to spring	1.0 lb/day to calving, 2.0 lb/day to spring
No. heifers	21	21
Avg. gain, lb	-142	-1
Avg. hay intake, lb/day	11.4	18.4

*From Lusby and Armbruster, 1976. Winter supplementation not substitution. Oklahoma State Univ.

HOW TO SUPPLEMENT

Protein supplements are available in many forms. Cottonseed cake has been the old standby, but cubes containing non-protein nitrogen, protein blocks, and liquid feeds have also become very popular.

Cottonseed cake is still one of the best, and is used by a large number of stockmen. Similar in form is the range cube, a product or products produced by innumerable feed companies and country mills. Range cubes are usually represented on a crude protein basis; e.g. 30% range cubes, 40% range cubes, etc. They almost always contain urea and are therefore usually cheaper on a protein basis than cottonseed cake. A small amount of urea can be used quite successfully (no more than 1/3 the crude protein equivalency in most cases), but the stockman should be wary of products containing more than 1/3 the crude protein equivalency from urea. As explained in Chapter 12, range cattle can utilize only relatively small amounts of urea. Likewise, many range cube products contain grain byproduct fillers. As explained earlier, starch is not of particular value in range supplementation. There are, however, a large number of quality products available which are primarily oilseed meals with some added urea. In addition, the better products will usually contain some extra phosphorous and trace minerals, as dry forage is always lacking in phosphorous, and usually low in trace minerals. [Cottonseed meal by itself is also quite high in P (1.0-1.25%)]. The addition of Rumensin, a feed additive cleared for pasture feeding, would also normally be a good addition to range cubes (page 75).

The only drawback to cottonseed cake or good quality range cubes is that they require hand feeding. As with feeding grain, this will cause cattle to quit grazing early and possibly eat less forage.

Figure 2-6. When cows are hand fed a supplement they will typically quit grazing early and will wait to be fed their supplement. Feeding every other day tends to reduce this problem.

In order to reduce the amount of labor required to hand feed cake or range cubes, many ranchers have gone to feeding twice as much every other day. Experiments done by extension services and several universities have compared every day with every other day feeding of cottonseed cake, and results have been comparable (Melton and Riggs, 1964). Care should be exercised in the use of supplements containing urea, however, as every other day feeding will reduce the amount of urea that can be utilized (see Chapter 12).

Figure 2-7. An advantage to protein blocks is that when properly formulated and managed, cattle do not "hang around" the feeding area. Placed near water, cattle will come up and consume a portion of the block, and then return to grazing. This ensures maximum utilization of the pasture, which should be the cheapest feed available.

The use of protein blocks is another popular way of supplementation. The big advantage to the blocks is that when properly used, they are self feeding. Salt, other minerals, special binders, sometimes unpalatable ingredients such as ammonium sulfate as well as the degree of hardness in the block itself are used to control consumption. Reputable companies will usually market several different blocks which vary in hardness and palatability, so as to be able to more closely control consumption under different conditions. As with most manufactured goods, there are some inferior products on the market. Unlike range cubes, high urea contents are not usually the problem with cheap block products since if urea levels get too high, the block form will not hold together. Rather, poor quality block products will usually contain relatively large amounts of grain and grain by-products which make the blocks quite palatable, and thereby precludes self feeding.

With good quality blocks, consumption can usually be regulated within .2 or .3 lb. Thus, as much as 10 days to 2 weeks supply may be put out at one time. The blocks are usually placed near water, so when cattle come up to drink, they can lick on the blocks for a while, and then go on back to grazing. Cattle do not seem to hang around protein blocks like they do self limited grain mixes.

In the past decade or so liquid feeds have become very popular. Using molasses as the base, and urea or other forms of NPN as the "protein" source, these products are fed in either open troughs or troughs containing lick wheel arrangements. Phosphoric acid is usually added to regulate viscosity, as well as to add phosphorous to the mix. In addition, phosphoric acid is also often used as a palatability regulator. Some products also contain trace minerals, although settling out is often a problem.

In the author's opinion, liquid feeds are usually the least desirable way to supplement protein to range cattle. While they work reasonably well in high grain rations, the amount of non-protein nitrogen contained in liquid feeds is frequently in excess of what range cattle can utilize (see Chap. 12). The reasons for the popularity of liquid feeds on rangeland is that the cost per ton is usually much less than the cost of dry supplements, they are convenient to use, and cattle take to them quite readily.

SPECIAL MANAGEMENT PROBLEMS ASSOCIATED WITH LARGE RANCHES

The most serious, complex, and difficult problems the ranch manager must deal with are personnel management problems. Technical production problems are usually identifiable and often have scientifically developed solutions. The only problem is obtaining the information necessary to correct or avoid the situation. Agricultural extension and research services were developed for the specific purpose of disseminating such information.

To the man actually managing a large ranch, however, the greatest problems he faces do not deal with the agricultural sciences. He faces personnel management problems of a kind and intensity unknown to ordinary business managers. Not only must he cope with the usual personality conflicts, etc. that arise when men work together, but he must also deal with the additional problems intensified by the fact that they must live together as well (at the ranch headquarters). The managers of large ranches not only must be personnel managers, but also landlords and civil mediators. Essentially they become the mayor and chief of police of small villages. Getting up at 2:00 or 3:00 AM to quell a marital dispute is all part of a ranch manager's job.

Due to the costs of bringing in power and water lines, telephones, etc., most ranch headquarters will have a number of family dwellings placed quite close together. In the author's opinion this is a real mistake. Not only must the men get along, but under these conditions, their families must get along as well. When wives are feuding it is extremely difficult for their husbands to remain aloof. Even simple quarrels between children can precipitate problems. The men may be able to get along while working, but when they go home they are forced to take sides.

In ranches with a fairly large turnover of people, a marital infidelity will often develop. When this happens the manager's job can literally become a nightmare. It's none of his business, but he is still often expected to step in and take some kind of action. Obviously, there is no easy solution or course of action.

For these reasons, it is the author's recommendation that employee housing be placed as far apart as possible. If housing is already in place or if other considerations require the close proximity of housing, large walls or fences should be built between the units. Anything that contributes to privacy will be of help. Also, the housing should be as similar as possible in every detail. It is amazing how jealous employees (particularly their wives) can become if they feel that an employee of equal rank has a nicer house or mobile home. Even things as seemingly trivial as individual appliances can precipitate problems.

PERSONNEL MANAGEMENT IN RANCH MANAGEMENT

Large ranches can have as many as 20 to 40 employees, and yet formal personnel management principles are seldom, if ever, practiced. However, there is a definite need for some basic personnel management structuring.

The most commonly violated principle of personnel management is that of simple planning and communication. In most businesses, employees are called periodically into informal meetings where the plans and goals as developed by the management are discussed. The intent is to let each employee know what his function is and what is expected of him.

Very seldom do ranches hold such meetings. The various ranch foremen may know approximately what the manager wants done, but the average ranch employee doesn't have any idea what he is going to be doing from one day to the next. It is commonplace for a man to get on a horse or into a pickup in the early morning and follow a foreman out, not knowing where they are going or what they are to do once they get there.

The second most violated principle of personnel management is the doctrine of Unity of Command. This simply means that each employee should take orders from only one superior. The idea is to eliminate confusion as to what the employee is to do.

Because of the large geographical area that ranches cover, and the subsequent physical communication problems that result, the principle of Unity of Command should be rigidly adhered to. In many instances, however, it is not. When it isn't, not only does confusion exist, but quite often serious personality conflicts develop.

The situation is that most large ranches will have a manager and two or more foremen. One foreman will tell a man to do something, and then later another foreman or the manager will come along and tell the man to do something else. The man will then attempt to carry out the last order. In the meantime, the first foreman will come back to check on the man and either find him gone or doing something else. His first thoughts are usually anger, as the man has disobeyed him. If he reprimands the man (as is often the case), the man will obviously be resentful.

By holding meetings each employee is better able to understand just what he is supposed to do. Even such tasks as gathering cattle would benefit by a short meeting. Quite often cowboys aren't even told where the cattle are to be moved to, or the route to be taken. Thus, they must divert some of their attention to see which way the foreman appears to be going. If the ranch is brushy or hilly, this really creates a problem, and cattle inevitably are missed. To the uninitiated, this may sound like an exaggeration, but it is not. Ranch foremen are typically stoical people, and usually treat questions as to what is to be done with incomplete, vague answers, and in some cases, even contempt.

Communication not only eliminates confusion, but also bolsters employee morale. It is human nature to want to feel important, and keeping someone in the dark can only make them feel insignificant. Someone who feels that they are being used as a pawn certainly isn't going to take as much pride in their work as someone who feels that they are appreciated.

THE MOST COMMON SINGLE MISTAKE MADE BY RANCH MANAGERS

The biggest mistake made by ranch managers is usually trying to work their men too hard. In all of agriculture the protestant work ethic is very much in evidence. Ranchers and farmers take great pride in their operations and consider working 7 days a week, 12 hours a day, a normal procedure. Because long hours are literally part of their lifestyle, they think nothing of working their employees an equal number of hours. When an employee is unwilling or disenchanted with working such long hours, they often think of him as lazy . . . someone not cut out for ranch work, someone who needs one of those soft, 5 day a week town jobs.

What they don't realize is that their employees don't have the pride of ownership, or in the case of the hired manager, the bonus incentive or social prestige* that drives them to dedicate their every waking hour to the ranch. They don't realize that the employee is only trying to make a living for himself and his family. He therefore cannot be expected to have the same drive as those involved in management.

When employees are pushed that hard they will either quit or become less conscientious in their duties. It is not at all uncommon to see large ranches with 80 - 100% yearly employee turnover rates. There are, of course, some good men that will stay under those working conditions, but as a general rule, the type of man that stays is not the best employee. Rational men cannot be expected to devote their life to a business that they have no vested interest in.

The author cannot overemphasize what a serious problem this is. Jobs that require skill or experience (calving out heifers, checking heat, operation or repair of machinery, etc.) will not be performed as well as they should be, either because of lack of experience (due to turnover), or because of lack of concern (employees pushed to their limit simply cannot function as well as they would ordinarily). The noticeable results are abused vehicles, machinery, and other equipment. Less noticeable are lower calf crops (due to less conscientious insemination, calving assistance, etc.) greater calf mortality (due to improper vaccination, treatment, etc.), and even larceny (due to contempt for the management).

*In small western communities managing a large ranch is a very prestigious position.

CHAPTER 3   COW-CALF FARMING

Many of the principles which apply to western cow-calf ranching also apply to what will be described as cow-calf farming. For this reason the reader should review the sections on cow-calf vs. stocker operations (page 19), spring vs. fall calving (page 20), supplementation (page 29), early weaning (page 28), creep feeding (page 25), and the yearly production cycle (page 21). In addition, it would also be a good idea to review the section on range management (page 10).

This type of cow-calf production generally applies to the type of production commonly practiced in the Midwest and most of the Southeast (except Florida). In addition, there are areas in the western U.S., such as the Willamette Valley of Oregon, which also practice this type of cow-calf production.

There are a few operations that run breeding cows as the main enterprise. These are usually purebred operations as land and other production costs in these areas are usually much greater than the return from commercial cattle will afford. For the most part cow-calf production is usually incidental to farming.

Cows are typically used to graze wooded or hilly areas that cannot be farmed, and to clean up crop residues. Marginal fields are often seeded to pasture grass for additional grazing. Cows are also sometimes grazed on winter wheat fields, or other small grain pastures.

Calves from the cow herds are either sold or kept for use as stockers for further grazing, or as growing cattle for silage feeding programs. As yearlings, the cattle are often further held for fattening programs.

THE GRAZING & FEEDING CYCLE

The actual grazing and feeding cycles of these operations vary considerably, but as a general rule, begin with the cattle grazing the wooded lots or improved pastures during the spring and summer. In the Midwest the improved pastures may consist of a number of pasture grasses such as bluegrass, fescue, orchardgrass, one of the bermuda grasses, etc., whatever is best adapted to the area. In addition, these pastures may also be seeded with a legume such as birdsfoot trefoil or one of the clovers. Having the grass-legume mixture increases the length of time grazing is available as the legumes usually grow well during the heat of summer when the grasses are often dormant. Likewise, during the cooler periods of the early spring and fall, the grass usually supplies most of the grazing.

After harvest in the fall, the cows are often turned into the fields to glean crop residues which can be used by beef cows. The main considerations for using these materials are just to see to it that the cows are properly supplemented (page 77), and that they are moved whenever most of the better material is gone.

After the crop residues are gone, the cow herd is usually either put on small grain pastures, or moved into drylot for winter feeding of harvested materials. Sometimes a combination of both is used. Silage is probably the most common winter feed, although cornstalks, straws, hays, and haylages are also used.

SPECIAL CONSIDERATIONS

GRAZING UNTILLABLE PASTURES

Essentially the same range management principles as outlined on page 10, apply. The only difference being that these areas typically have higher rainfalls, and thus can withstand somewhat more grazing pressure than the more arid western lands.

GRAZING IMPROVED PASTURES

The types of grasses used in improved pastures are usually more productive and capable of withstanding much more grazing pressure than native species. As a consequence, they usually also require more water and fertilization. Actual utilization will depend upon the variety used.

As with range management, improved pastures are usually more productive if some sort of grazing system is used; i.e. concentrating a large number of the animals in a portion of the pasture or pastures for a short period of time, rather than grazing the entire pasture area continuously. If there are several pastures available, this is accomplished simply by moving the animals periodically. In the case of large pastures, this can be accomplished by confining the animals to one section of it via electric fences.

Since improved pastures are usually fertilized*, the stockman should be aware of the potential for nitrate toxicity (see page 81). Likewise, bloat (page 78) can also be a serious problem if the pasture contains considerable legumes, or lush green grasses.

GRASS TETANY— On lush green pastures (primarily small grain pastures) one of the biggest problems is what is known as Grass

---

*Fertilization of improved pastures usually only consists of N, P, and K. Seldom, if ever, are trace minerals used in fertilizer. Since cattle continuously remove trace minerals by grazing, but none are put back in the soil, it is a good policy to supplement trace minerals.

Tetany, Magnesium Tetany or sometimes Wheat Pasture Poisoning. For a yet unexplained reason periods of rapid forage growth accompanied by cloudy, cool weather will cause the magnesium content of the grass to become unavailable to the animal. Normal magnesium levels are present by chemical analysis, but some unknown compound or compounds tie it up and make it unavailable. Magnesium is required for normal muscle contraction, and so when it is unavailable to the animal, a state of tetany develops. Cows nursing calves are particularly susceptible since the magnesium requirement of lactating cows is two to four times that of dry cows and stocker cattle.

The first signs of the disorder are overall nervousness, and blank staring. As the condition proceeds, the skin may begin to twitch (particularly on the face). The final stage is usually violent convulsions. If treatment is not administered promptly, the animal will usually die.

Treatment usually consists of intravenous administration of calcium-magnesium gluconate (calcium shares many roles with magnesium in muscle contraction). Only persons with experience in veterinary medicine should give this treatment, because if given too rapidly or in too great a quantity, the heart will fibrillate and the cow will die. Unexperienced persons can give a subcutaneous injection of 200 cc of saturated magnesium sulfate solution which will increase blood magnesium levels in about 15 minutes (USDA, 1975).

Incidences of grass tetany can appear very quickly, and so the stockman may not be physically able to treat his cattle in time. This places a special emphasis on prevention, which means getting some supplementary Mg into the cattle. There are a number of methods to do this, such as spraying Mg compounds directly on the fields, adding Mg to the water, etc. The easiest and by far the most popular method is to feed a mineral mixture with an extra high level of Mg. The drawback to this method is being sure the cattle eat adequate amounts of the mineral mix. To ensure consumption, many feed companies either add or recommend adding a palatability enhancer (soybean meal, grain, etc.) to the mix. As further protection, it is the author's recommendation to feed dry hay to pasture cattle during periods of rapid forage growth. Likewise, anytime one animal appears to have grass tetany, the entire herd should be hayed that day.

GRAZING CROP RESIDUES

The grazing of crop residues is discussed at length on page 76, which deals with stocker cattle. The primary difference between

Figure 3-1. Cows grazing a stalkfield. When run on stalkfields cattle will eat all the grain (lost during harvest) first, and will therefore gain weight.

Figure 3-2. When all the grain in a stalkfield has been consumed, and cattle are forced to eat the stalks and leaves, they will begin to lose weight. In addition, because the concentration of nitrates is greatest in the portion of the stalk closest to the ground, the danger of nitrate toxicity increases when the only available forage is the stalks themselves.

grazing stocker cattle and grazing cows is that cows can be run on poorer quality residues, since weight gain is not necessarily required. This means that cows may be left in the fields longer than stocker cattle. As explained on page 77, most fields will contain a considerable amount of unharvested crop (lost during combining, etc.). Cattle will clean up the concentrate first (corn, sugar beets, etc.) and then when forced to, will eat the residue itself.

The important thing to remember is that cattle cannot maintain their bodyweight on residue itself (cornstalks, wheatstraw, etc.). They will lose weight when all they have is residue. The typical pattern is that a farmer puts a herd of cows on a cornstalk field, etc., and removes them 60 to 90 days later. When removed, they may weigh the same as they did when placed on the fields, but they did not maintain their weight on cornstalks. During the first half or two-thirds of the period they gained weight by cleaning up what corn, etc. was available, and lost weight during the last part of the grazing period when the concentrate was gone.

As explained on page 29, supplementary protein must be provided for cattle fed crop residues (low quality roughages). Residues are usually low in phosphorous and trace minerals, and so a mineral supplement must also be provided. In the Midwest, stalkfields can be extremely low in Mg, which can create the same symptoms of tetany normally attributable to grass pastures. In problem areas increased Mg must be supplied.

As also mentioned in the section on stocker cattle, crop residues should be tested for nitrates before turning cattle in. This is even more important for cows since they will be left in the fields longer; i.e. nitrates are contained in the residue itself and since cows will be forced to eat a greater amount of the actual residue, the danger is greatly intensified. In fields with high nitrates the typical pattern is that cattle will do fine and appear healthy for the first few weeks. As the amount of concentrate decreases and the cattle are forced to eat larger quantities of the residue itself, nitrate toxicity will begin to appear, often quite rapidly.

CONFINEMENT FEEDING AS A COMPLETE COW-CALF PRODUCTION SYSTEM

In the past, the feeding of cows in drylot for the purpose of raising calves has seldom been profitable. This doesn't mean that there is anything inherently wrong with confinement feeding, only that cattle prices have not been sufficiently high to cover the costs. The big cost of course, is feed. In most years, just the cost of the feed for the cow has been as much as the return from a 400 lb calf:

Hay @ $80/ton ($.04/lb)

16.9 lb/day during dry period x 160 days = 2,704 lb
25.4 lb/day during lactation x 205 days = 5,207 lb
                                           7,911 lb

or $316.44*

Silage @ $25/ton ($.0125/lb)

39 lb/day during dry period x 160 days = 6,240 lb
58 lb/day during lactation x 205 days =11,890 lb
                                        18,130 lb

or $226.62**

*400 lb calf would have to sell for $.79/lb ($\frac{316.44}{400 \text{ lb}}$) just to cover cow feed costs.

**400 lb calf would have to sell for $.57/lb ($\frac{226.62}{400 \text{ lb}}$) just to cover cow feed costs.

The actual physical act of producing calves from confined cows presents no particular problem. It requires more management than present systems, but it is feasible. Indeed, there have been a number of demonstration projects conducted at various state universities; Oregon (Price, 1977), South Dakota (Deutscher, 1975), Arizona (Taylor, 1975), Michigan (Ritchie, 1975), Purdue (Perry, 1974), Texas (Marion and Riggs, 1972), Alabama (Harris, 1970), and Illinois (Albert, 1969).

Confinement feeding has been researched to provide basic production information should it ever be needed. At the present time there is growing pressure from urban preservation/ecology groups to eliminate grazing from public lands. Public lands make up over 50% of the western states and support a large percentage of the nation's beef cow herd. If these cattle were removed, confinement feeding could be used to make up some of the shortage. However, the economics would still have to be right. Either feed prices would have to come down, or cattle prices would have to go up.

The author does not want to sound so presumptuous as to predict economic trends, but if the past is any indication of the future, it seems apparent that feed costs will not come down far enough (relative to cattle prices) to permit confinement feeding. Prices may come down substantially in individual years, but it seems unlikely that they could stay down long enough to permit long term investment in cow-calf confinement facilities. If cow-calf confinement becomes feasible, it will most likely be due to an increase in cattle prices.

Once it does become economically feasible to produce calves in drylot, investment in facilities should come, as the cost on a per head basis is a fraction of what investment in a ranch or pasture system would be. Ritchie (1975) built a confinement lot for 100 cows at a per cow cost of $265. Investment in ranch property would run upwards of $1,500 per cow unit.

USE OF "WASTE" ROUGHAGES
Much has been written about the many high cellulose waste products that could be used as cattle feed. Examples would be cornstalks, various straws, sawdust, cotton gin trash, etc., etc. Each of these materials can be treated chemically to enhance the digestiblity, and theoretically could be used to support beef cow herds. The most successful treatments have involved the use of sodium hydroxide or acid hydrolysis under steam pressure.

The big problem to date has been cost. The materials (straw, etc.) are sometimes available for the price of harvesting, but to date the cost of chemical treatment has elevated the cost of the finished product to equal or exceed the price of conventional roughages. In addition, these products are usually quite low in vitamins, minerals, and digestible protein. This calls for supplementation that is usually in excess of what would be required for conventional feeds.

There are many who contend that at some point in time the use of chemically treated "wastes" will be mandated due to expanding human populations. Certainly, lowly productive animals such as beef cows would be the logical livestock to be fed these products. Much more research will be required, however, before the use of these products can be advocated for the year round maintenance of a cow herd. To date, most of the research has involved short term trials. Typically the treated product consisted of only a portion of the ration. Numerous problems could develop with long term feeding, which may not be readily apparent with short term observations. For this reason the producer/investor should be very cautious concerning developing an operation around the availability of these materials.

USE OF BYPRODUCT FEEDS

As mentioned on page 105, there are a number of byproduct feeds available from time to time (cull fruits and vegetables, etc.). The products with a relatively high value are usually bought up and incorporated in feedlot and dairy rations. The other products are usually very low in digestiblity, and so essentially the same situation as mentioned under "Waste Roughages" (previous section) prevails.

MANAGEMENT PROBLEMS ASSOCIATED WITH CONFINED COW-CALF PRODUCTION

Aside from economics, there are a number of problems associated with cow-calf confinement. The most commonly reported problem has been calf disease, bacterial scours in particular (Ritchie, 1973; McGinty, 1972; Slyter, 1970; and Albert, 1969). Outbreaks of calf diseases would certainly be affected by cleanliness of the pens, weather, etc. Something that can also precipitate calf disease is the fact that during calving, calves in confinement have been noted to suckle cows other than their dams (Price, 1977). Oftentimes these are cows that have not yet calved, but have begun secreting colostrum. E. coli are the organisms most often involved in bacterial scours, and it has been shown that colostral antibodies are required for the prevention of E. coli infections (Radostits, 1975; Logan, 1974; and Dam, 1972). To ensure that each calf gets its full portion of colostrum, pregnant cows should be kept separate from cows that have already calved.

The second biggest problem associated with cow-calf confinement is seeing to it that each cow gets her share of feed. In a feedlot this creates no problem since the cattle are fed ad libitum (all they will eat). This, of course, is because it is desired that the animals gain weight. However, with a brood cow the intent is only that she maintain her weight. This means that with most feeds, the cows must be limit fed; i.e. given a predetermined amount rather than all they will eat. Cows will establish a pecking order rather quickly, which can result in the more timid cows receiving as little as 50 to 60% of the feed allotted them (see Figures 3-3, 3-4 & 3-5). In other words, if silage is being fed at 35 lb/cow, a ration of 3,500 lb would be fed to a pen of 100 cows. Under most conditions what would happen is that 15 or 20 of the most timid cows would only get about 20 lb.

Figure 3-3. Typical pecking order behavior pattern of confined cows fed a maintenance ration. Note the size of the brisket of the intimidated cow compared to the cow she is fleeing. Note also the hip bones of the intimidated cow versus the cow at the feed manger.

Bunk space per animal must therefore be increased considerably (over normal feedlot standards). In feedlots, 9 to 12 inches of bunk space per animal is usually considered adequate, but for cow-calf confinement feeding, the author would recommend an absolute minimum of three feet. Still, pecking order problems may persist as aggressive cows can intimidate more timid cows 10 to 15 feet away. The most timid cows may not even come to the feedbunk until the other cows have had their fill and moved away.

In feedlots, a common practice is to feed twice a day to ensure fresh feed. For cow-calf confinement, a better practice is to feed every other day. By feeding twice as much every other day, the "boss" cows aren't able to monopolize the feed quite so much. They eat their fill, but it isn't as great a percentage of the total feed available, leaving more in the bunk for the timid cows. Under most conditions two days feed are eaten in one day.

This procedure may not be necessary, however, if waste roughages are used. In that case high fiber may reduce intake to the point of making ad libitum feeding necessary.

Figures 3-4 & 3-5. Pictures of cows involved in 2 yr long confinement study conducted at Oregon State Univ. Cow at top was low in the social order of a group of cows receiving 100% the NRC maintenance requirements. Cow at bottom was high in the social order of a group of cows receiving only 88% the NRC maintenance requirements. Both cows started the experiment in moderate flesh. (Price, 1977)

GENERAL MANAGEMENT RECOMMENDATIONS FOR COW-CALF CONFINEMENT

At the time of writing, there were no commercial year round cow-calf confinement lots in operation (to the best of the author's knowledge). The reader should therefore keep in mind that the recommendations given here are derived from research projects, rather than commercial production.

PEN REQUIREMENTS

Pens for cows and calves should provide no less than 300 square feet per cow-calf pair. If muddy conditions develop, that space should be doubled or even tripled. Also some type of shelter should be available. It is highly desirable that the flooring in the shelter be concrete, to facilitate cleaning. When the calves are young, clean bedding should be made available in the sheltered area.

As mentioned on page 48, a minimum of 3 feet of bunk space per cow-calf pair should be provided. In addition, an area of bunk space inaccessible to the cows, should be provided for the calves. This will allow them access to some of the roughage, as otherwise, the cows will monopolize it. Likewise this area can be used to feed a higher quality ration to the calves if creep feeding is desired.

Ordinary feedlot water troughs are suitable for the cows, but are quite often too tall for young calves. Waterers should therefore be set lower, or two waterers at different heights should be used.

BREEDING PROGRAM

Because the cows are confined, artificial insemination would seem to be very applicable. Not only would heat detection be greatly simplified, but calving problems associated with the typically larger birthweights produced by superior A.I. sires could be spotted much more readily.

Natural service obviously requires bulls to be kept in the pens (unless hand mating were practiced). When kept in confinement, bulls inevitably do some type of damage to the facilities. Calves are usually from two to four months old during the breeding season, and bulls will occasionally cripple or kill calves in confinement with them. If natural service is used, a higher bull to cow ratio can be utilized than under pasture conditions, particularly range conditions.

In the last few years there has been much discussion in the scientific press concerning the efficiency of small cows versus large cows. The theory being that they eat less forage and therefore stocking rates can be increased (under pasture conditions). Their calves are smaller, but there are a greater number of them (due to the higher stocking rate).

In confinement the reduced requirements for the smaller cow are very apparent. According to NRC (1976), an 882 lb beef cow requires about 7.2 lb of TDN during pregnancy and 10.4 lb during lactation. A cow weighing 1,102 lb, requires 8.6 lb TDN during pregnancy and 11.7 lb during lactation. On a yearly basis (205 day lactation), the 882 lb cow needs 3,284 lb TDN, and the 1,102 lb cow needs 3,774 lb TDN; a difference of 490lb TDN.

If silage contains 24.5% TDN, then an additional 2,002 lb will be required. In essence, an extra ton. At $25 per ton for silage and 60¢/lb for calves, the calf from the larger cow must be more than 41 lb heavier at weaning to pay for the extra feed consumed by its dam.

In a study conducted by Texas A & M, large (1048 lb) and small (865 lb) type Hereford cows were compared under semi-confined conditions. When bred to a Brown Swiss bull, the calves weaned from the large Herefords were only 6 lb heavier than the calves from the small Herefords (458 vs. 452 lb) (Hammack, 1969).

John Riggs, who has been responsible for overseeing the 12 year long confinement experiment conducted in Texas, has said that the Angus x Jersey cow is probably the ideal cow for confined calf production.* They are small, consume less feed than larger cows, but produce as much or more milk than most beef cows. Jerseys are noted for easy calving, and thus the little "Ajax" can be bred to large terminal sires (Charolois, Simmental, etc.).

In one study (Thomas, 1971), Angus x Jersey (A x J) cows were compared to Hereford and Charolois (C) cows in confinement. Results are shown in Table 3-1. The A x J cows ate about 25% less feed than the Charolois cows, yet produced calves only 7% smaller. In feedlot studies after weaning, the C x AJ calves were only 3% less efficient than the C x C calves.

In a commercial operation obtaining A x J cows would be difficult and probably expensive (considering the discount that A x J steers would bring). The point here is that the size and subsequent amount of feed consumed by the cows should definitely be considered. The cost of feed and the sale price of calves will determine what size cow should be used. If feed is high relative to the price of calves, the smaller cow should be considered. If feed prices are low compared to calf prices, the larger cow may be more profitable.

*1972 Personal Communication.

Table 3-1. WEIGHTS, PRODUCTION, AND FEED CONSUMPTION OF CONFINED COWS.

Breed		Weight of cow	Ave. milk yield	Weaning weight	Ave. Feed Consumption	
Sire	Dam				Dry	Lact.
Char.	A x J	697 lb	14.7 lb	510 lb	12.7 lb	20.8 lb
Hereford	Hereford	1109	12.3	468	14.1	23.3
Char.	Char.	1258	14.4	547	17.0	27.1

WEANING PROGRAM

It would seem that early weaning would be advantageous under most confinement situations. The reason being that it is inherently more efficient for a calf to consume and convert feed into growth directly, rather than the cow consuming the feed, converting it into milk, and then the calf consuming the milk.

As mentioned earlier, the cow's nutritional requirements increase greatly with lactation. Early weaning obviously cuts down on the amount of cow feed required, by reducing the amount of time a lactation ration needs to be fed.

CHAPTER 4   BACKGROUNDING

Backgrounding is a term that has come into use in the cattle business only relatively recently. It refers to the handling and treatment of freshly weaned calves so as to get them over their stress and health problems, before they are placed on feedlot rations or turned out to pasture.

In most cases the backgrounding operation is a stockyard which specializes in "straightening out" freshly weaned calves. The cattle are held for a three to four week period, and then are shipped to their ultimate destination.

This facet of the cattle industry essentially began with the simultaneous emergence of the cattle feeding industry, and the "importation" of calves from the Southeast. The feedlot industry developed in these areas because of relatively mild climates, and availability of feed. Other areas in the western and midwestern U.S. still have viable feeding industries, but by far the bulk of cattle are fed in the areas outlined in Figure 4-1. Such great concentrations of cattle obviously required the movement of cattle from outside areas. The majority of these cattle have come from the Southeast.

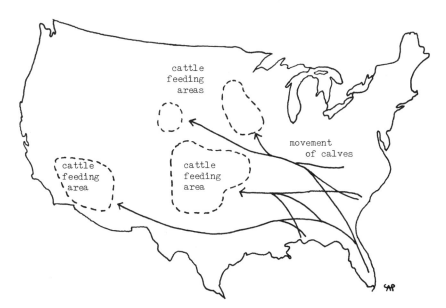

Figure 4-1.   UNITED STATES FEEDER CATTLE MOVEMENTS

Moving cattle those great distances created substantial stress. Coupled with the fact that the calves were usually freshly weaned, substantial health and death loss problems developed. The term "shipping fever" came into use as a description of the respiratory/ pneumonia disease complex which came to be associated with shipped in cattle. Shipping fever currently causes more loss to the cattle feeding industry than any other physical production problem.

At times the losses can be absolutely catostrophic. Death losses have run as high as 25%; 10% losses are not uncommon.

Figure 4-2.   Classic shipping fever symptoms.

The average feedlot recognized that it did not have the resources to cope with such a problem. As a result, the feedlots either created a special division within their own organization for "straightening out" calves, or farmed them out to backgrounding operations.

DISCUSSION OF SHIPPING FEVER AS IT USUALLY OCCURS

The situation begins with a calf nursing a cow, several hundred miles away from his ultimate destination. While nursing the cow, the calf gets an extremely high quality protein (milk). Likewise, milk is high in available carbohydrates and fats; i.e. it is a good source of energy. Then suddenly one day the calf is cut off the cow and shipped to a salebarn or stockyards. There, almost invariably the calves are given grass hay as the only feedstuff. Being freshly weaned, it will be a while before the calves will begin to eat anything. For a period of time, they will simply bawl for their mammas. The amount of time will vary, but this behavior will usually last from 6 to 24 hr.

The longer the calves stay off feed the weaker they become, but even when they do begin to eat the situation is still far from good. Young calves need a feed with a protein content of from 12-14%, but grass hay typically contains only 5-10%. Obviously the calves suffer a protein deficiency. Then too, whenever ruminants do not eat for a period of time, the microbial population in the rumen decreases. Reduced rumen microbial populations mean reduced utilization of cellulose. Since cellulose is the principal energy source in grass hay, the calves do not receive the energy that they should. All the hay does is fill them up.

The calves are then loaded onto trucks and shipped. Transit time will usually vary from 24 to 72 hr. While in transit, they certainly don't eat anything, and rumen microbial populations are reduced that much more.

Upon arrival at their final destination, the usual feed provided is again grass hay. Grass hay is used at auctions and stockyards because it is usually relatively cheap, and at feedlots because it will not scour the calves or create digestive upsets. However, the fact remains that grass hay does not provide enough protein or energy. The protein deficiency created is particularly detrimental since it reduces the ability to form leukocytes and antibodies.

Incoming calves are often treated with antibiotics which aid in the prevention of bacterial diseases, but not viral diseases. Antibiotics are ineffective against viruses. Bacteria and viruses are two totally different types of organisms, and it is of utmost importance that this be understood. Bacteria are relatively large organisms. Viruses are infinitesimally smaller. So small, that they cannot be seen with a microscope. Viruses are visible only with an electron micrograph (a device capable of making individual molecules visible). There are no commercially available drugs capable of affecting viral infections.* The only way an animal can combat a viral infection is through its own internal defense mechanisms. When a virus invades an animal's body, antibodies are made to combat the virus. Antibodies are specific for each type of virus, and are essential to the animal's resistance to disease.

---

*At the time of writing (1981), research into interferons had begun. These compounds have a potential for being used as anti-viral agents, but it will apparently be many years before they are made available for commercial veterinary use.

Antibodies are primarily composed of gamma globulin, a complex protein. When a calf is undergoing a protein deficiency (fed only grass hay), obviously it cannot form antibodies at an optimum rate.

This is especially important in shipping fever, since it is generally recognized to be a two stage infection, initiated by a virus. A viral infection creates a lesion within the respiratory tract, which creates an avenue for a secondary bacterial infection. Parainfluenza $_3$ (PI$_3$), Infectious Bovine Rhinotracheitis (IBR), and Bovine Virus Diarrhea (BVD) are the viruses most commonly involved. Bacteria of the Pastuerella family are the most common bacteria involved, although Haemopholus somnus has also been identified recently.

In outline form, Shipping Fever would be represented by Table 4-1.

1. Virus - PI$_3$, IBR, BVD (primary infection)
2. Bacteria - Pasteurella (secondary infection)

TABLE 4-1.   SHIPPING FEVER.

The viral infection is seldom fatal in itself. Its primary effect is weakening the animal and creating a lesion in the respiratory tract which leaves the animal susceptible to the secondary bacterial infection that actually kills the animal.

COURSE OF THE DISEASE

The calves arrive tired, stressed and frightened. They receive poor quality rations, which are typically low in energy and usually provide either not enough protein or low quality protein. Being low on energy the calves must use their body stores for energy, and/or catabolize what protein is available to them for use as energy. At the same time they come into contact with a variety of viral diseases.

If the feedlot or ranch cowboys are observant, the sick calves are noticed and "pulled" for treatment. Treatment usually consists of antibiotic injections which serve to protect against the secondary bacterial infection. If the calves are not pulled in time, the secondary bacterial infection occurs, and the calf dies.

The calves that are pulled and treated in time, often become known as "chronics", or calves that never get well. They don't die, but they never recover and go on to gain weight or even look healthy. While some chronics defy explanation, the cause of many chronics is simple nutrition. Quite often, sick pen rations consist of nothing more than hay. Other times, the ration will contain a protein supplement which contains substantial amounts of urea, or natural proteins from a low quality source (designed for the fattening ration). Thus the animal continues to undergo a protein deficiency and therefore cannot form antibodies in sufficient amounts to cope with the viral infection. Daily treatment with antibiotics prevents the lethal bacterial pneumonia from occurring, but inability to combat the sublethal viral infection continues, and so the animal remains a chronic.

NUTRITION OF SICK & STRESSED CATTLE

Figure 4-3. It takes more than just antibiotics to make sick cattle well again.

The nutrition or ration considerations for sick or stressed cattle are substantially different than for healthy animals. This is primarily because sick and stressed animals usually have altered digestion patterns.

Sick animals typically go off feed, and stressed animals which have been shipped ordinarily do not have the opportunity to eat for a period of time. This obviously limits the amount of nutrients in the rumen, which likewise limits the ability of the microbial population to maintain itself. As a result, the microbial population will be reduced in proportion to the length of the fasting period.

Microbial populations of shipped in cattle can be reduced to as low as 10% that of a normal animal.

Reduced microbial populations obviously means reduced rumen digestion since there are less organisms present to break down and utilize the feed. In addition, sick animals also usually have reduced rumen motility, which further reduces rumen fermentation.

Reduced rumen fermentation (digestion) limits the ability of the animal to utilize fibrous materials and to upgrade proteins; i.e. take proteins unbalanced for amino acids and reform them (via microbial action) into balanced proteins that the animal can use. Reduced rumen fermentation also means that a larger proportion of the ingested feed will be passed to the lower tract (for digestion). Figure 4-4 explains the situation diagramatically.

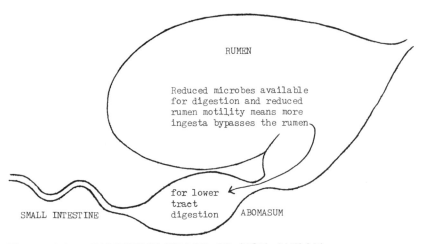

Figure 4-4.   DIGESTIVE TRACT OF SICK ANIMAL

Reduced digestion in the rumen, and increased digestion in the lower tract, changes ration considerations (for sick cattle) in three ways.

1. The level of urea or other non-protein nitrogen compounds must be reduced.

2. Amino acid balance of protein supplements must be taken into consideration.

3. Cellulose cannot be relied upon as a primary source of energy, and therefore readily available forms of energy such as starch (from grains), must be fed.

USE OF UREA FOR SICK CATTLE

Because sick cattle have reduced digestion in the rumen, they cannot use the same level of urea or other forms of NPN that normal cattle can, since rumen fermentation is required for the utilization of urea. In addition, approximately 10 to 14 days are required for cattle to become adapted to urea. Obviously, shipped in calves that have recently been weaned off pasture probably haven't been fed any feeds containing urea, and thus won't be adapted.

If cattle have been fed urea for a period of time and are adapted, it should be remembered that urea adaptations can be lost relatively quickly. Therefore, even if cattle have been adapted, once they become sick and go off feed for a period of time, they lose their adaptation.

Feeding urea to sick cattle in what would be moderate levels to healthy cattle causes two problems. The first problem is that significant portions of urea will be either passed to the lower tract, or lost as NH_3 through the rumen wall (see Chapter 12). Thus it will not be utilized and will be processed by the liver and kidneys for excretion. This probably places a certain amount of stress on the animal. Obviously it is unwise to place any additional stress on an already sick animal. The second problem is that the urea is supposedly there to meet part of the animal's protein requirement. If it can't be utilized, then quite clearly it cannot supply the animal with the protein it needs.

AMINO ACID CONSIDERATIONS FOR SICK CATTLE

In terms of protein utilization by ruminants, amino acid balance is commonly considered to have no significance. Sick cattle are a definite exception. Depending upon the condition of the animal, it can be of critical importance. Extremely sick animals will have almost no rumen motility and if off feed for a substantial period of time, they essentially become monogastric animals.

SOURCES OF ENERGY FOR SICK CATTLE

Because of the reduced rumen fermentation, sick cattle cannot utilize roughages to the extent that healthy cattle can. This is contrary to the practices of many feedlots which provide hay only or high roughage rations to sick cattle.

Since rumen fermentation is reduced, cellulose (the major energy source in roughages) cannot be relied upon as a primary energy source for sick cattle. Readily available carbohydrates must be provided. In most cases, this means some type of grain must be fed.

Figure 4-5.   A common problem in many feedlots is to feed only hay
to the sick pen. Cattle will readily consume hay, but it does not
supply the nutrients they need to fight infection.

This is very important because if sufficient utilizable energy
is not supplied in the ration, vitally needed protein will be deam-
inated (broken down) to be used as an energy source. In addition,
fat reserves will be used in an attempt to supply energy. Weak
and sick cattle often have little fat reserves and so muscle tissue
is then broken down to yield energy. This obviously further weak-
ens and debilitates the animal.

RATION CONSIDERATION FOR STRESSED OR SICK CATTLE
The biggest obstacle to overcome in keeping stressed cattle
from becoming sick is to get them eating quickly. Likewise, if
sick cattle are to recover, they must begin eating again.

As mentioned in the previous section, rations for stressed
or sick cattle must contain readily available energy (grain), be
low in urea, and supply protein from a high quality natural source.
To say that the ration must be palatable, understates the case;
the cattle or calves must eat the ration. This may sound simple,
but it is not. Getting stressed cattle to eat requires strict attention
to detail, and quite literally determines whether or not a backgrounding operation will be successful.

In designing and presenting a proper ration, there are nearly
as many "don'ts" as there are "dos". To begin with, do not use

fermented feeds . . . no silage, haylage, or high moisture grains. The biggest reason is that fermented feeds are not readily palatable to cattle, calves in particular. It takes a period of time for cattle to develop an appetite for fermented feeds. They may eat a little bit at first, but nothing (in terms of dry matter consumption), like they will eat with palatable dry feeds. The second reason for not using fermented feeds is that their utilization requires ruminant digestion, silages and haylages in particular. Again, stressed cattle are often not fully functioning as ruminants.

Dry rolled or ground grains should be avoided whenever possible. The big reason being to reduce dust. Stressed animals, many of which have contracted a respiratory disease, certainly do not want a dusty feed. If dry rolled or ground grains cannot be avoided, molasses must be added to control dust. Molasses does increase the palatability somewhat, but not to the extent that is commonly believed. Its main contribution (to palatability) is simply to control dust.

Probably the only time dry processing of grains may be necessary is when milo is the only grain available. If corn is available, it is far better to feed it whole, than to process it. Calves will do a very thorough job of chewing corn, and will eat considerably more corn when it is fed whole than if it is dry processed. If oats are available and are reasonably priced, they make an excellent starter feed for calves. Oftentimes sick cattle will eat oats when they will eat no other grain. As with corn, it is best to feed oats whole.

Dry chopped or ground hay should also be avoided. Again, the reason is to avoid dust. The author would prefer to use long hay, and top dress a concentrate mix over the hay, or use pelleted hay in the mix, or both. Alfalfa and other legume hay should usually be avoided, as it has a tendency to scour stressed cattle. Grass or cereal hay is a much better choice.

Cottonseed hulls are a byproduct readily available in most of the major cattle feeding areas. They have very little nutritive value, but for some reason cattle seem to crave them. When possible the author generally recommends adding 10 to 15% cottonseed hulls to a starter ration. They give an attractive odor to the ration and the lint tends to hold fines and dust. Calves will usually start on a ration with cottonseed hulls quicker than rations without them.

When selecting a protein supplement, quality is of the utmost importance. The author would recommend using a soybean meal base. Combinations with cottonseed meal have worked quite well in the past. Whether additions of animal sources of protein would be of benefit is difficult to say as there has been very little experimental work like that done. It is the author's undocumented opinion, however, that there would not be a measurable benefit. Urea can be added, but only at extremely low levels.

When purchasing all materials, the foremost consideration must be quality. Hay must be clean, and absolutely free of mold or mustiness. Likewise it must be bright, green, and free of weeds. One must be particularly careful when purchasing cereal hays as they mature quickly and are often cut too late. Grains must be clean and free of dust and fines. If pelleted roughages or supplements are used, the pellets must be clean and hard - free of fines. If cottonseed hulls are used, they must be clean and fresh . . . hulls from piles left on the ground for a period of time will have collected dirt and will not have the characteristic odor that cattle find appealing.

To purchase high quality, clean feed materials, one must be prepared to pay more. The difference in price per unit of material may seem substantial, but the difference in cost must be compared against death loss.

BASIC IMMUNOLOGY

While veterinary medicine is beyond the scope of this text, the author believes that an understanding of the principles of immunology is basic to setting up or understanding backgrounding programs.

Figure 4-6.   Lack of immunity can be the result of stress, or poor nutrition, as well as a breakdown in the vaccination program. Indeed, a vaccination program cannot succeed if proper nutrition is lacking, and excessive stress can do much to overide everything.

Whenever a pathogenic organism gains entrance to the body, an immunological reaction is triggered. The first line of defense is the white blood corpuscles or leukocytes. In addition, antibodies, which are specific for the invading organism are produced. Once produced, the antibodies identify and destroy the invading organism they have been produced to protect against.

Some pathogens produce toxins which cause the animal distress, rather than the organism itself. In that case, the body will also produce an antitoxin which is used to neutralize the effect of the toxin. Examples would be tetanus and overeating disease (Clostridium perfringins).

How seriously a pathogen will affect the animal depends upon the ability of the animal to build antibodies or antitoxins versus the severity of the exposure (infection). The condition of the animal at the time of the exposure makes a great difference in the ability to respond to an infection. Stressed, frightened, undernourished, shipped in calves are obviously in poor condition to fight off infection.

If the animal has been exposed to the disease previously, there will usually be a number of antibodies already present in the bloodstream. In some cases the level of antibodies present is great enough to prevent any noticeable infection. This is the idea behind vaccination.

Vaccination against a disease involves giving the animal a weakened or killed form of the disease-producing bacteria or virus. In the case of diseases that are caused by toxins, a weakened form of the toxin is given. The animal then forms antibodies or antitoxin in response to the introduction of the vaccine. Thus, when the animal is later exposed to a naturally occurring virulent form of the disease, the immunity (antibodies/antitoxin) it developed in response to the vaccine is hopefully great enough to prevent infection.

Vaccines using weakened pathogens are called attenuated or modified live vaccines. They are developed by weakening the bacteria or virus in some way. Other vaccines simply use killed pathogens. Vaccines using killed pathogens seldom cause vaccinated animals any distress. Modified live vaccines, however, can stress vaccinated animals, particularly if they are in poor condition.

Modified live and killed vaccines produce what is known as active immunity. That is, the animal itself produces antibodies in response to the vaccine.

There is also what is known as passive immunity where antibodies from a donor animal are introduced into another animal.

An example of this would be a calf receiving colostrum. The calf's mother secretes antibodies into the colostrum and when the calf drinks it, he is given passive immunity by the antibodies he receives. There are some vaccines which are made by taking antibodies from hyperimmunized animals. The antitoxins are often of this type. With some diseases, such as tetanus, both an antitoxin and a toxoid are available. The antitoxin is made of serum taken from animals that have been exposed to tetanus many times. The toxoid consists of denatured toxin. The antitoxin produces passive immunity, and the toxoid provides active immunity.

Active immunity is a much more longlasting immunity since the animal itself actually produces the antibodies. Passive immunity lasts for only a short time period; i.e. only as long as the injected antibodies remain viable. The advantage to passive immunity is that it provides immediate protection.

FACTORS TO CONSIDER IN VACCINATING SHIPPED IN CATTLE

Ten days to two weeks are usually required before a calf can develop immunity from a vaccination. Almost coincidentally, most feedlot diseases apparently have a 10 day to 2 week incubation.

This obviously means that cattle should be vaccinated prior to coming into the feedlot; ideally, 2 to 3 weeks before weaning. Under the marketing conditions that exist today that would be virtually impossible to accomplish on any kind of a commercial basis. The vast majority of the cow-calf producers are simply not going to do it. There has been a move, particularly in the Midwest to try and accomplish this, but the problem is certifying that the vaccines were given correctly. Cow-calf producers are notorious for exposing vaccines to sunlight, heat, etc. For these reasons buyers have been and will probably continue to be cautious about paying premiums for specially treated cattle.

A dilemma is created by the fact that immune response and incubation period for most feedlot diseases are approximately the same length of time. The problem is that many of the vaccines are modified live products. Since live vaccines can actually stress the animal to a degree, theoretically they can contribute to a more severe infection of the disease in question. (If exposure to the virulent organism occurs at about the same time as the vaccination; i.e. upon arrival at the feedlot.) Killed vaccines, of course, do not create this type of problem.

Another problem the backgrounding operation must be concerned with is the condition of the calves at the time of vaccination.

If the calves are stressed and weak, they may not be capable of responding to the vaccine. This is particularly true if they are undernourished. As explained several times earlier, an animal cannot form antibodies to fight a disease if it does not have the nutrients (protein primarily) to form antibodies. Likewise, antibodies cannot be formed in response to a vaccination if the nutrients required are not present.

HANDLING VACCINES

Time and again there have been outbreaks of diseases in cattle that were vaccinated against those particular diseases. With shipped in cattle the problem is often the result of the cattle not having the physical or nutritional well being to be able to respond to the vaccine (as explained in the previous section).

Oftentimes, the outbreak is due to simple mishandling of the vaccines used. The most commonly made mistake is setting the vaccine bottle in the sunlight. Modified live vaccines can often be completely inactivated after only a few minutes in direct sunlight. Even if the vaccine is not completely inactivated, the potency will still be reduced.

The next most commonly made mistake is allowing modified live vaccines to come into contact with heat. Oftentimes stockmen are very careful to keep their vaccines refrigerated until the time of use, only to take them out and set them by a branding fire, or up on the dash of a pickup (where heat and light are intensified coming through the windshield).

Progressive operators will usually keep their vaccine in some sort of ice chest during use. This not only keeps them cool (chests should be filled with ice), but also protects them from sunlight (the only time the lid is open is when the bottles are used to fill syringes). Feedlots will often have refrigerators within a few feet of the processing facilities.

Many vaccines must be mixed with other solutions for activation. Potential problems can occur if excessive amounts are mixed at one time. The typical situation is enough is mixed at one time for a whole day's run, or more than is required is mixed up for one pen or bunch of cattle. In the case of the latter, the natural reaction is to want to save the excess for later use. A similar situation occurs when the correct amount is mixed up, but the crew breaks for lunch, etc. Usually these vaccines must be used immediately after mixing. Letting them sit for a length of time reduces their effectiveness, particularly at room temperature.

Another form of mishandling often seen is the mixing of vaccines and other medicines. Some vaccines can be mixed with other products, but many cannot. For the sake of convenience (giving one shot instead of two or more), far too many operators are guilty of mixing vaccines and other medications without checking to find out if the products can be mixed.

Handling and administering vaccines and other medications correctly is relatively simple. All one need do is read the instructions on the product to be used. Aside from that, one must remember that vaccines are relatively unstable compounds. The label will always read vaccine, but if handled incorrectly, the contents will not be vaccine.

Figure 4-7. The label on a bottle of vaccine will always read "vaccine". But if it has not been handled properly the contents will not be vaccine.

CHAPTER 5   STOCKER OPERATIONS

Stocker operations are usually defined as taking calves (250 - 550 lb) and putting them on some sort of pasture for sale later on as yearlings (600 - 850 lb). Most operations will usually consist of either: (1.) Spring and summer grazing of native or improved pastures; (2.) Winter grazing of cereal pastures (primarily wheat); or (3.) Fall and winter utilization of stalkfields (cornfield residue primarily).

One of the main reasons many operators choose stocker over cow-calf is the reduced labor requirement. Once the cattle have been properly backgrounded the only major labor requirement is keeping the cattle scattered evenly or rotated over the grazing area, and periodically checking on health. Upkeep on facilities such as fences, corrals, etc. is often reduced since stocker cattle usually do not do the damage that breeding cattle do. There is not the labor requirement for calving assistance, branding and vaccinating, etc. demanded by cow-calf operations.

Stocker operations are usually thought of as having a higher risk than conventional cow-calf ranching. This is because the purchase price of the stocker calf is the major cost input in the yearling calf that is sold 3-8 months later. This puts the stocker operator at the mercy of the cattle market. If the market goes down (after he bought his calves), he may lose money regardless of how efficiently his cattle gained weight.

For example, assume the stocker operator bought 400 lb calves for 60¢/lb to sell as 650 lb yearlings. If his cattle gained at 35¢ per lb, this means that final sale price will have to be $327.50 [(400 x $.60) + (250 x $.35)]. The costs attributed to his own direct management account for only 27% of the final sale price ($87.50 ÷ $327.50). The other 73% were dictated to him by the market, something over which he has no control.

At the present time the progressive stocker operator may not need to expose himself to this kind of market risk. The feedlot industry is becoming increasingly active in forward contracting, and this form of marketing is usually available for the operator who is willing to spend the time and effort to seek out a contract buyer.

In addition to forward contracting, feeder cattle may now be sold on the Chicago Mercantile Exchange futures market. Selling on the futures market requires some sophistication and knowledge (page 153), but is available to operators willing to study and learn about this form of marketing. However, the futures market should not be considered by those unwilling to spend the time required to become familiar with the mechanics of it.

STOCKER OPERATIONS ON NATIVE PASTURES

This type of operation is normally limited to spring and summer grazing as that is the only time when native pastures are good enough to support weight gain. Indeed, there are some areas in the western U.S. that will not afford reasonable weight gains (at least 1.0 lb/day) even at these times, and so stocker cattle are not a viable option to some ranchers. Breeding cattle need only maintain their weight, and such areas are better suited for that use.

The fiber content of the forage essentially determines whether the cattle will gain weight or not. As explained on page 195, fiber is difficult to digest and so not only does it reduce the nutritive value of the forage, but also reduces the amount that the animal can eat. Thus, poor quality forage has a snowballing effect on reducing gain.

Table 5-1 shows the amount of TDN required for 441 lb steers to maintain or gain weight.

Gain	Lb TDN Required
0	4.2
1.1 lb	7.5
1.5 lb	7.9
2.0 lb	8.1
2.4 lb	8.6

Table 5-1. TDN REQUIRED FOR GAIN OF 441 LB STEERS.*

*From NRC 1976 (Appendix Table II)

Calves weighing about 450 lb pastured on lush green native grass with a crude fiber content of 20-22%, will eat approximately 2.75% their bodyweight in dry matter. Consumption will therefore be about 12.3 lb of dry matter. Lush grass of a native variety can have a TDN value of approximately 62-66%. Thus, TDN consumption will be from 7.6 lb to 8.1 lb. Looking at Table 5-1, it can be seen that these cattle will gain from about 1.2 to 2.0 lb/day.

A 450 lb calf run on dry, mature forage with a crude fiber content of 35-40%, however, will only be able to eat about 2.0% its bodyweight in dry matter (and only then with proper supplementation). Grass with that much fiber will only have a TDN value of from about 47 to about 52%. Thus, dry matter consumption will be about 9.0 lb and TDN consumption will range between

4.3 and 4.7 lb. Looking at Table 5-1 it can be seen that the cattle will gain only .1 to .2 lb.

If pasture costs are 40¢/hd day, then cost of gain for the cattle on lush grass would be from $.20 to $.33/lb. Cost of gain for the second group of cattle would be from $2.00 to $4.00/lb.

These examples are extreme, but not altogether uncommon. Every year there are large numbers of stocker cattle run on pasture that would be better suited for breeding herds. Even more common is the practice of holding stocker cattle in pastures too long; i.e. after the forage has become dry and mature due to the onset of hot or cold weather, or drouth.

During the peak grazing months, these pastures will often support gains of up to 2.0 lb/day. Once the heat or drouth of summer appears, or in other areas the first frost of fall arrives, the forage quality declines rapidly; animal gains drop off just as rapidly. As mentioned earlier, a very common management error is to see stocker cattle left out too long. Over the grazing season many operators will experience average gains of 1.0-1.25 lb/day. If they were to move their cattle off 30 to 40 days sooner, many would experience average gains of 1.5 to 1.75 lb/day.

Obviously then, one should be certain that his pasture or ranch is capable of producing the amount of gain required of stocker cattle before undertaking that type of operation. Many knowledgeable cattlemen state a minimum of 1.0 lb/day is required before a stocker operation can be considered.

Knowing when a pasture or ranch is capable of being used in a stocker operation is relatively simple. In most cases all one need do is look to see if large quantities of green grass are available. Arriving at proper stocking rates takes some expertise, but good advice can usually be obtained from the SCS, range extension specialists, or private consultants.

STOCKER OPERATIONS UTILIZING IMPROVED PASTURES

Improved pastures, both irrigated and dry land, are used all over the U.S. for stocker cattle (small grain pastures are discussed in a later section). Their use usually stems from the fact that they are of marginal value for crop farming, but can support more productive cover crops than native grasses.

When good quality cropland is diverted for use as pastureland, it is usually done to complement a more extensive cattle operation. Some farmers, however, have chosen to put cropland into pasture since it requires less investment in machinery (no harvesting equipment), and less labor.

Still, the majority of the pastureland used for stocker cattle is land that for some reason is not quite suitable for crops. It may be too hilly, rocky, have too shallow a soil profile, etc., etc. But when planted to improved pasture grasses or legumes, its value is enhanced (compared to its value with only native grasses).

In the southeastern U.S. many improved pastures are utilized for cow-calf operations. This is a fairly common practice in sub-tropical and tropical areas since the pasture grasses which are adapted to tropical climates are usually of relatively low nutritive value. With ample fertilization they produce copious amounts of tall, rapidly growing forage which can support a relatively high stocking rate of cows.

In more temperate climates, improved pastures usually produce forage that is too good in quality for efficient utilization by cow herds; i.e. it is only desired that beef cows maintain their weight, whereas improved pastures usually permit gains of from 1.0 to 2.0 lb/day. From an efficiency of production standpoint good quality pastures should be utilized by stocker cattle or dairy cows as they are more productive animals. Productive means that they are able to produce more edible protein (meat, milk) on the same amount of forage. That is, a beef cow must graze all year to produce one 300-500 lb calf. Stocker cattle, however, will gain 250-350 lb during a 5-7 month grazing period, and will consume less than half as much feed. Cows will, of course, gain weight (along with their calves) on high quality forage, but if the cow is to be bred back, the extra weight is of no real advantage and can be a disadvantage at calving time.

In most years the economics of production parallels the efficiency of production and so most improved pastures (in temperate regions) are used for stocker cattle. Those operations that do use improved pastures for cow herds are usually registered operations, although some commercial operations will use improved pastures during the breeding season.

AGRONOMIC MANAGEMENT OF IMPROVED PASTURES

The initiation and optimum management of improved pastures is a very localized type of thing. Actual recommendations will depend upon local weather and soil conditions. Thus, almost each individual farm will require an individual production plan. For this reason only general principles will be covered here (rather than specific recommendations).

FORAGE SPECIES USED

The first consideration in most improved pasture systems is

to provide grazing for the longest period of time possible. To accomplish this, two or more species are usually required. Oftentimes cool season grasses are mixed with legumes. The grasses, such as fescue or rye grow well during the cooler months of spring and fall, and legumes, such as clover or vetch, will grow during the hot summer months. In addition, the nitrogen fixing properties of legumes decreases fertilization requirements. Again, the actual species chosen will depend upon a myriad of factors pertaining to local conditions.

GRAZING MANAGEMENT OF IMPROVED PASTURES

Grazing management of improved pastures is similar to range management (see page 10), the major difference being that improved pastures can usually withstand much greater grazing pressure, particularly if irrigated. When grazed extremely close, most pasture species will recover when ample moisture and fertilizer are provided. However, during the regrowth phase, most species require a period of time before grazing should be permitted again. As explained on page 11, this is because plant energy reserves are utilized during regrowth. If the leaf area of the plant is continually removed before the plant can carry on photosynthesis and replenish the reserves, then the plant will die.

For best results, a rotational grazing system is usually best. Cattle are concentrated in one pasture or section of a pasture until the desired amount of available forage is utilized. The cattle are then moved to another pasture or section of pasture, and the previously grazed area is given a chance to recover. The use of electric fences greatly facilitates this practice. In the case of large pastures, the wires are simply moved up and down the pasture.

If cattle are left to graze one area over an extended length of time (as in continuous grazing), spot grazing occurs because cattle will tend to select plant regrowth rather than more mature growth. Continuous grazing therefore does not usually result in as good of utilization as rotational grazing. Gain per animal will be greater with continuous grazing since the cattle can select the most lush (highest quality) forage, whereas gain per acre will be greater with rotational grazing since the forage is utilized more completely.

STOCKER OPERATIONS ON SMALL GRAIN PASTURES

Winter small grain pastures (wheat, barley, oats, and rye) make exceptionally good grazing and are used extensively for stocker cattle in all but the more northerly states. Many small grain fields are used specifically for grazing, but by far the majority are used for both grazing and grain production.

Figure 5-1. Cattle grazing winter wheat. With proper management, grazing actually enhances grain yields.

With proper management grazing does not reduce grain production. Indeed, proper grazing actually enhances grain production. Periodically cutting off part of the blade or leaf area of the plant (grass) tends to make the roots spread out and send up more shoots (blades of grass). The plant actually becomes hardier. An analogy would be mowing a lawn. Most lawn grasses respond to mowing with increased growth and density (assuming proper fertilization and irrigation).

Small grains are, of course, annual plants. Planting usually takes place in August or September and grazing will normally start in late October to November. If grain is to be harvested, the cattle must be removed before mid-spring (before the boot stage); grain harvest normally takes place in early summer.

Small grains (wheat in particular) have a tremendous regrowth ability, and can withstand extremely heavy grazing pressure. However, this great regrowth ability can cause problems in the form of bloat.

Whenever weather conditions have stifled growth (extreme cold, cloud cover, etc.), and then very favorable weather conditions appear, bloat can be a substantial problem.

Crude protein of cereal grasses can run as high as 30% and TDN as high as 78%. Normal ranges would be 20-28% CP and 68-75% TDN. Gain is usually limited to 1.5-2.0 lb/day, even though the TDN content is relatively high. This is due to the high moisture content (72-78%) which limits dry matter intake*, and the high protein content which causes it to be quite laxative.

SUPPLEMENTATION ON SMALL GRAIN PASTURE

Since small grain grasses are quite high in protein, protein supplementation is not needed. Mineral supplements are needed since small grain pastures are seldom fertilized with anything except N, P, and K. In addition to trace elements, mineral mixes on small grain pastures should also be relatively high in magnesium, calcium, and phosphorous.

FEEDING GRAIN ON SMALL GRAIN PASTURES

Feeding grain on pasture is a rather common practice. Sometimes it is done to increase gain, but more often the intent is to "stretch" the pasture. Grain feeding will usually accomplish both those objectives, but there is a question of efficiency.

Normally 4 to about 6 lb of grain are fed daily to pasture cattle. Gains will ordinarily increase about .5 lb. Cattle that would have gained 1.5 lb/day will therefore gain about 2.0 lb/day. Thus, the conversion of grain to extra gain normally runs from 8:1 (4 lb ÷ .5) to 12:1 (6 lb ÷ .5).** In a feedlot those conversions would be considered quite poor for cattle of stocker weight (400-700 lb), as conversions of 5.5 to 7:1** would be more typical of feedlot performance.

The question becomes, "How much less pasture grass did the grain-fed calves eat?". If they ate substantially less grass, then it can be concluded that grain feeding is reasonably efficient. Unfortunately, there is very little detailed research in this area. Armbruster of Oklahoma State University has been involved in a number of extension service projects and reports that empirical observations of grain on grass projects show grain/extra gain conversions to run from 8 to 16:1. His general recommendations are that if the pastures are well utilized, then grain feeding is usually economically feasible, but if pastures are not utilized

*Cattle must eat approximately 3½ times as much small grain grasses as they would a dry feed (90% DM) to get the same dry matter intake.

** Air dry basis.

well (grass left over at the end of the season), then grain feeding can create some rather high cost of gains.

In other words the pasture grass is a relatively cheap feed, and the grain is relatively expensive. In this light, one can understand Armbruster's recommendation. If an operator is overstocked, or wants to increase stocking rate, he can feed some grain without hurting his overall cost of gain too much. However, if he feeds grain and is still understocked, overall cost of gain will increase considerably, since he did not fully utilize the cheap source of feed . . . the pasture.

The theory behind the poor utilization of grain on grass is essentially the same as explained on page 32. The energy contained in the grass is primarily in the form of cellulose whereas the energy contained in the grain is in the form of starch. Cellulose and starch require two completely different types of microorganisms for digestion; cellulolytic and amylolytic. When cattle graze for a period of time, the rumen microorganism population tends to be cellulolytic, but when the cattle then eat a big slug of grain, the pH of the rumen drops, and the population shifts to amylolytic bacteria. As a result, the rumen stays in a constant flux so there is never a good stable population of either type of microorganism. As a result, digestion of both feeds is impaired.

Along with the question of efficiency, there are some physical problems associated with feeding grain on grass. The first problem is limiting the grain. Unless the grain is fed in limited amounts, or an intake inhibitor is mixed with the grain, most cattle will go on a full feed of grain. This is usually undesirable as stocker cattle will become fat at normally unmerchantable weights, and the pastures will not get utilized.

Hand feeding is the simplist way of limiting intake. The main disadvantage is the obviously high labor requirement. In addition, the more aggressive animals will usually get a greater share of the grain than the more timid animals.

Most stockmen prefer to use a self feeder with some sort of intake inhibitor. Salt is usually used as the inhibitor. When salt is added to a grain mix at the rate of 6 to 10%, grain consumption of stocker cattle can usually be limited to 4 to 6 lb.

Having a self feeder in one part of the field is most certainly going to create uneven grazing. This may not be of consequence in fields to be grazed out anyway, but can make a big difference in fields to be harvested later for grain, as plant vigor in the area surrounding the feeder will be greatly reduced (from increased grazing pressure and trampling). In addition, the salt content of the manure deposited on the field will be quite high. The final drawback to feeding grain on pasture is that the more cattle gain during the stocker or growing phase, the less they will gain during a later feedlot or finishing phase.

Figure 5-2. One problem with self feeding grain on pasture is that cattle will spend considerable time lounging around the feeder. This creates uneven grazing and trampling.

USE OF RUMENSIN ON PASTURE

At the time of writing, Rumensin, the trade name for monensin sodium, a feed additive, had just been cleared by the FDA for pasture use. Rumensin reduces the methane loss during fermentation, and thereby makes rumen fermentation more efficient. In high grain feedlot rations, animal gain stays the same with Rumensin, but the cattle eat less feed, and thus have a better feed conversion (lb feed/lb gain). With pasture or high roughage rations, intake remains the same, but gain increases. The response to Rumensin on pasture appears to be related to forage quality. The lower the quality, the greater the response.

At the time of this writing the clearance for Rumensin is 200 mg/hd/day. Gains on dry grass will normally be increased 10-15%; e.g. from .75 lb/day to .85 lb/day, and gains on lush pasture will be increased approximately 2-3%; e.g. from 2.0 to 2.05 lb.

STOCKER OPERATIONS UTILIZING CROP RESIDUES

Figure 5-3. Stocker cattle grazing a corn stalk field. Cattle will gain weight on the grain left in the field, not the stalks. They will eat the grain first and only when it is gone will they begin eating substantial amounts of stalks. When that occurs, cattle will begin to lose weight. Therefore, when grain stops appearing in the manure, it is time to move the cattle.

Running cattle on crop residue pastures is a paradox for many cattlemen. Some cattlemen get excellent gains (up to 2 lb/day) while other cattlemen experience very poor performance on the same type of residue pastures.

The difference lies in the understanding of a few basic facts concerning residue grazing. The most important fact being that cattle cannot gain weight on crop residue (cornstalks, straw, etc.) . . . cattle gain weight by eating the wasted crop (grain, etc.) left in the field after harvest.

Of the commonly grown crops, cornstalk fields usually support the greatest weight gains. Typically, 3 to 8% of the corn grain is lost during harvest, and it is the grain, not the stalks, that the cattle gain weight on. Cattle are remarkably adept at rooting through stalks and leaves to sort out the ears lost during harvest. Likewise, cattle (calves in particular) are able to chew and utilize the corn quite well.

Knowledgeable cattlemen watch their cattle closely and when corn fails to appear in the manure, the cattle are moved. At this point the fields will not appear to have been well utilized as there will be a considerable amount of stalks, leaves, etc. left on the ground. This is because cattle will first seek out the leftover grain, and will eat the stalks only when forced to.

If cattle are allowed to remain in the field after the grain has been removed, gain will decrease rapidly. All too often the situation is that cattle will gain well for a period (while utilizing the available grain), and will later lose weight for a period (when forced to eat stalks). The end result usually being that the owner will conclude the stalkfield grazing doesn't work, as overall gains will typically be rather poor.

Other grain crops do not allow as much weight gain as corn. This is primarily due to the inability of cattle to chew whole wheat, milo, etc. (compared to corn). In addition, much of this grain is shattered and difficult for cattle to pick up off the ground (rather than being neatly contained on a cob). For the same amount of grain left in the field, one can only expect 40 to about 70% the gains of cattle on corn vs. milo, wheat, or barley.

Specialty crops, vegetables etc., are quite variable. Performance will vary with the individual crop and the maturity of the crop; i.e. often vegetable crops are pastured because adverse weather or other problems prevented harvest and the crop became too mature for sale as produce. Overly mature vegetable crops will not support the same performance that an equal amount of wasted vegetable at the correct stage of maturity. As with other forage this is because of an increase in fiber and a decrease in soluble sugars and carbohydrates. Some specialty crops, such as onions, contain alkaloids which can create problems if they comprise a large part of a cattle diet.

SUPPLEMENTATION ON STALKFIELDS

Cattle pastured on stalkfields should always be supplemented with protein, vitamins, and minerals as grain and stalks are deficient.

Vitamin A supplementation is an absolute must for calves on stalkfields as there is no appreciable vit. A or carotene activity in either stalks or grain. Old time cattlemen are often of the opinion that calves do not do well on stalkfields, whereas cows can. The rationale behind this thinking is usually that cows can utilize roughage better than calves. The truth of the matter is that calves actually do better on stalkfields as they are better able to sort out and chew the available grain. However, in the past calves have not done well on stalkfields because either cottonseed cake or other protein feeds devoid of vit. A activity were the only supplements provided. Cows were able to get by since they typically have enough liver stores of vitamin A to carry them 3-5 months. Calves typically do not have nearly as great a vitamin A store.

Calcium, phosphorous, and trace mineral content of stalk-field forage is very low. Mineral supplementation is therefore required. This may be accomplished via the inclusion of a mineral "pack" in the protein supplement, or the use of free choice mineral feeding.

LEASE AGREEMENTS ON STALKFIELDS

A preferred practice among many cattlemen is to lease stalk-fields on a per hundredweight/month basis; e.g. at $2.00/cwt. the cattleman would pay $10.00/month for each 500 lb calf on the field. Pricing stalkfields in this manner eliminates the guess-work in figuring how much grain is left in the field. In this way, as soon as the cattle stop passing grain in the manure, the cattle owner knows it's time to move the cattle. He pays the landowner only for the length of time his cattle were actually in the fields.

COMMON PHYSIOLOGICAL PROBLEMS WITH STOCKER CATTLE

PASTURE BLOAT

By far the most common cause of death in stocker cattle on small grain pasture is bloat. It can occur at any time, but increased incidences can be expected whenever there is a period of rapid plant growth. The most dangerous time is when very favorable weather conditions appear after a period of unfavorable weather conditions; i.e., rapid plant growth after a period of dormancy. The longer the plant has been dormant and the more the sub-sequent conditions favor growth, the more dangerous the situation.

Whenever small grain grasses undergo rapid growth, the protein content of the regrowth will be quite high. A significant portion of the proteins produced will be in a soluble form. The actual cause of pasture bloat is not clearly understood, but it is believed that soluble proteins are at least partially responsible.

This type of protein is a very gooey material. An analogy would be Elmers glue, which is primarily composed of casein (milk protein), a soluble protein. A large volume of gas is produced during rumen fermentation which the animal must eructate (belch) off. With these gooey proteins floating in the rumen ingesta, the gas has a tendency to create a foam or froth when it bubbles up through the ingesta.

The presence of foam or froth in the rumen hinders the ability of the animal to eructate. This is a natural inhibitory mechanism which prevents particulate matter pneumonia. If the animal were to get this gooey foam in the esophagus (during eructation), there is a good chance that it would also get in the trachea. Thus when the animal inhaled, the foam could be sucked down into the lungs. With a proteinaceous coating inside the lungs, the animal would not be able to exchange oxygen correctly, and could die of pneumonia. A similar malady called lipoid pneumonia is an affliction of scuba divers who breath air that has been contaminated with lubricating oil (usually from the air compressor used to fill their tanks).

As the lesser of two evils then, the animal will swell up with the excess gas. The swelling will appear most evident on the left side (see Figure 6-18).

Once bloat occurs, the animal can die within a matter of minutes or hours. If found soon enough, simply forcing the animal to move around is often effective. If the bloating is severe, the preferred treatment is to drench the cattle with mineral or vegetable oil. This is accomplished by running a tube down the esophagus and pouring a pint to a quart of mineral or vegetable oil down the tube. The oil acts as a surfactant (compound that increases the surface tension), which tends to keep the foam from occurring. Only mineral or vegetable oil should be used; petroleum oils can be highly toxic.

In feedlot bloat, sometimes a tube is passed down the esophagus and the excess gas allowed to escape. This usually doesn't work too well with pasture bloat, as the foam or froth tends to block the tube.

When the animal is in extreme distress and there is not time to drench with mineral oil, the rumen may be punctured to release the gas pressure. This should only be done as a last resort. The puncture should only be made on the left side, just behind the ribs. Indeed, this is where the swelling will be the most noticeable, since there are no organs between the hide and the rumen in this area. On the right side, the liver is in the way. The instrument used to make the puncture is known as a trocar. Essentially it is a large pointed and sharpened spike, contained within a barrel or cannula. The entire unit (spike and cannula) is driven into the rumen. When the spike is removed, the cannula remains in place, and gas escapes through the opening. The cannula may be left in place as long as necessary to continue to release gas.

The trocar is used instead of a knife because it provides a tube for the gas to pass through. An ordinary knife wound often allows a small part of the gas and ingesta to gain entrance into the abdominal cavity; i.e. get between the skin and the rumen.

Rumen ingesta is teeming with bacteria, and so a severe infection usually occurs.

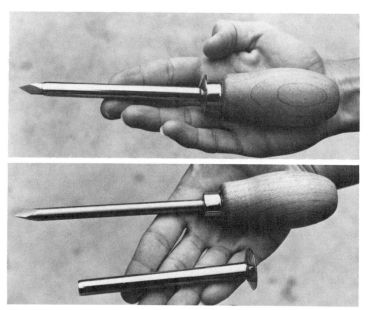

Figure 5-4. Trocar, an instrument only to be used as a last resort.

Regardless of whether the wound was made by a knife or a trocar, it will be very slow in healing. Gas will continue to push through and thereby greatly hinder healing. Peritonitis (infection within the gut cavity) is an ever present danger. Obviously then, puncturing the rumen should only be done when there is no time for any other type of treatment.

Prevention of a malady is often said to be more important than the cure. In the case of pasture bloat, prevention can be quite difficult. There is a commercially available surfactant known as Poloxalene which has been shown to be effective in reducing bloat. The problem is getting pasture cattle to eat it when they need it most. Soybean meal, molasses, and other agents are normally added to enhance consumption. The problem is that bloat occurs most often during rapid growth (when soluble protein production is greatest), which is also when pasture grasses are the most palatable. In actual practice it has been extremely difficult to get pasture cattle to consume products containing Poloxalene when there is lush, green forage available. If the product is sprayed directly on the pasture, as is often done in Australia and New Zealand, it works quite effectively, Likewise, the direct spraying of mineral or vegetable oil has also been shown to be effective.

In the U.S., the direct spraying of a product on pastures is usually not feasible. Probably the only practical way to eliminate bloat is to keep plenty of dry hay on hand, and watch the weather and cattle very closely. Whenever good growing weather appears after an extended period of poor growing weather, pull the cattle off the fields and fill them with hay before returning them. The hay dilutes the effect of the lush grass and stimulates rumen motility (causes the rumen to "churn" the digesta more). Likewise, anytime a number of cattle suddenly show up with bloat, it is a good idea to hay the cattle.

NITRATE TOXICITY

Nitrate toxicity is a relatively common problem with cattle grazed on fertilized pastures or stalkfields. Nitrate toxicities can also occur in cattle fed silages or hay that is high in nitrates. Also there have been cases of cattle contracting nitrate toxicities from contaminated water supplies. Ground water can be contaminated by seepage from manure or sewage, and surface water can be contaminated by runoff from fertilized fields or manure (from corrals, etc.).

The chemical formula of nitrate is NO_3, but when consumed by the animal, nitrate (NO_3) is converted in nitrite (NO_2) which actually does the damage. The NO_2 molecule displaces the oxygen normally carried by the hemoglobin in the blood. Adhering to the hemoglobin molecule very tightly, the NO_2 molecule will not allow the blood to transport oxygen normally. If the concentration of NO_2 is high enough, the animal will eventually suffocate.

The classic symptom of nitrate toxicity is blood that appears a chocolate brown color. This is due to the absence of oxygen in the hemoglobin and the presence of the NO_2 (methemoglobin). Other symptoms include poor performance and debilitation. If the concentration is high enough to cause death, it may come slowly or rather quickly.

The treatment for cattle with acute cases is intravenous injection of methylene blue, which acts as an oxygen carrier. Prevention of nitrate toxicity from harvested feeds and water is usually rather simple, since analyzing NO_3 is relatively inexpensive. If a feed is found to be relatively high in nitrates, it can be fed by diluting with other feeds known to be low in nitrates. (The rumen microorganisms can utilize a low level of nitrates as a source of nonprotein nitrogen.)

Pasture situations are more difficult to prevent since the concentration of nitrates can fluctuate. Plants transport nitrogen in the form of nitrate, and when growth of the plant is slowed excess nitrates may accumulate. Thus, whenever plant growth is retarded due to drouth, etc., and/or when heavy applications of N fertilizer are made, the cattleman should be wary. If pasture grasses have about 8,000 or more ppm nitrate (dry matter basis), the cattle should be moved. When normal growth of the plant is resumed, the nitrate level will decrease, and the cattle can be returned. Obviously this can create an extra inconvenience and expense to the cattleman, but death loss can be sudden and severe.

Nitrate toxicity occasionally occurs with stalkfield grazing. The greatest concentration of nitrates in plants is at the base of the stalk, and since stalk stubble makes up the majority of available forage, nitrate toxicity problems can become serious. When toxicities do occur, it is usually after the cattle have been on the stalkfield for a period of time. This is because cattle will seek out and eat the grain and leaves before they will eat the stalks.

Figure 5-5. A dangerous situation. Cattle being allowed to drink out of an irrigation water retention pond. Nitrates in tailwaters can often run high enough to cause sudden death losses.

A relatively common source of nitrates and subsequent toxicities is caused by allowing cattle access to irrigation tailwater. Rather than setting up water troughs and hauling water, some operators are tempted to allow the cattle to drink out of irrigation canals or tailwater pits. Once water has run across land that has been fertilized, it is almost certain to contain high levels of nitrates. Even if the concentration is not high enough to create a visible toxicity, it can reduce performance. Cattle should never be given access to tailwaters. Even if tested and proven safe, a good rain can fill pits or canals with enough runoff to create nitrate problems.

URINARY CALCULI (WATERBELLY)

Urinary calculi are caused by the precipitation of dietary minerals in the urine to form a stone-like object. In steers, where the size of the urethra is reduced due to castration, the stones can become lodged and block the flow of urine. This, of course, causes the animal great pain, and can eventually cause the bladder to rupture and kill the animal. There is a surgical procedure for saving the animal which is described on page 112. For urinary calculi occurring in pasture cattle, the mineral most often involved is silica. In areas where grasses contain high levels of silica, the problem can be acute. (In areas of New Mexico, the silica content of grasses can run as high as 20% of the ash. Many of the local ranchers in these areas will not winter steers just because of the waterbelly problem.)

There is no generally recognized and accepted preventative treatment. One of the foremost researchers on this subject, however, has communicated to the author that the feeding of an organic acid such as phosphoric acid or a cation such as magnesium oxide would be helpful.*

*Dr. Stan Smith, New Mexico State Univ., personal communication.

CHAPTER 6   CATTLE FEEDING

THE SOUTHWESTERN FEEDLOT

Figure 6-1. Typical southwestern feedlot. Felt, Feedyard, Felt, Oklahoma.

Between the late 1950's and the early 1970's, the cattle industry in the United States went through enormous changes. Prior to the 1950's, most of the grain-fed cattle came from the Midwest. The western part of the U.S. served primarily as a source of calves for the midwest farmer/feeder.

The creation of the Southwestern Feeding Belt was primarily due to development of hybrid grain sorghums (milo) which opened a vast new area of grain production. The inhabitants of this area (stretching from southern California, through Arizona and New Mexico, and up to the Texas and Oklahoma panhandles and parts of Kansas) were generally knowledgeable as cattlemen, and so a feedlot industry soon developed; i.e. they had ready access to both feed grains and cattle.

In the early 1960's the steam flaking method of grain processing was developed which gave most varieties of milo within 5-8% the energy value of corn. Dry heat processing of grain (popping and micronizing) which produced similar results, was developed soon after. Milo responds to heat processing much more than any other feed grain. Heat processed milo will usually have 7 to 12% greater digestible energy than dry processed milo. This meant that the feedlots that were large enough to be able to justify the expense of heat processing equipment, were able to substantially outperform the smaller lots that could not afford such equipment. This technological advantage brought about the demise of a large number of the smaller feedlots. The Southwestern feedlot industry came to be dominated by a relatively few large feedlots.

As the feedlots grew in size they outstripped the availability of capital obtained through normal channels. Their capital requirements exceeded the loan capacities of the local banks, and correspondent banks were extremely hard to come by. The large city banks weren't interested in corresponding on investments they knew nothing about several hundred miles away.

In their search for additional capital, the feedlots discovered that cattle feeding was an excellent avenue for tax shelter. The feedlots then became extremely active in the marketing of tax oriented cattle feeding investments. These investments were marketed through stock brokers in the form of a prospectus and were enormously successful. This caused the size of many feedlots to increase further, as marketing these securities cost in the neighborhood of $100,000 - $250,000. Expansion came very rapidly. Yards with one time capacities of 40,000-70,000 head were almost commonplace. At this point in time, the success of many of the large feedlots became more dependent upon their ability to market securities, rather than their ability to feed cattle competitively.

Just as tax oriented cattle feeding reached its peak, the infamous cattle market crash of 1974 occurred. Losses of over $100/hd on fed cattle were quite common. The attrition among investors and feedlots was high. At the time of writing, the feedlot industry had rebounded but not to the level of 1973. Due to unpleasant experiences in the 1974 crash and more stringent IRS rulings, tax shelter investment does not play the role in cattle feeding that it once played, but outside investment capital is still a major factor in southwestern cattle feeding.

All in all, cattle feeding in the Southwest has become a relatively sophisticated business. The typical large southwestern feedlot keeps consulting nutritionists and veterinarians on retainer, relies heavily upon urban, non-agriculture investors as feeding customers and sources of capital, and utilizes futures trading as a method of reducing risk exposure.

THE MIDWESTERN CATTLE FEEDER

As mentioned in the previous section, prior to the 1950's most

86

Figure 6-2. Confinement type feeding common to the Midwest (photo courtesy Steve Aldrich, Feedlot Magazine). This type of operation is used to cope with the severe cold during winter periods, as well as the mud that can be generated during spring rains and snow melt (Figure 6-3 below).

of the grain fed cattle came from the Midwest. The introduction of hybrid grain sorghums in the Southwest has now made that area the major cattle feeding area of the U.S.. The Midwest, however, still remains a factor in the U.S. cattle feeding industry.

Traditionally a farmer/feeder affair, the feedlot type of feeding is becoming more of a factor in the Midwest. However, the feedlots in the Midwest usually do not have capacities of over 2,000 head, and are often of a slatted floor/confinement design to enable them to cope with the severe winters encountered there. There is some commercial feeding, but much more commonly the cattle are owned by the feedlot operator. In most cases the feedlot is a secondary enterprise meant to complement a farming operation (provide a market for silage, high moisture grain, etc.).

CATTLE FLOW

The typical flow for steer cattle (Figure 6-4) is for them to:

1. Be bought as calves (300-500 lb) by an order buyer (often out of an auction ring).

2. Be sent to a backgrounding yard which holds them for approximately 30 days to get them over their inital stress and disease problems.

3. Be sent from the backgrounding yard to either a growing yard (which utilizes high roughage rations; to gain about 1.5-2.0 lb/day), or put out as stocker cattle on native or improved pastures, cereal pastures (wheat, etc.), or crop residue pastures.

4. When weighing from 600 to 800 lb be sent to a feedlot to gain about 3 lb/day.

5. When weighing 1,000 to 1,200 lb be sold as slaughter cattle.

6. Go to consumers as 600 to 800 lb carcasses.

There are, of course, exceptions to this flow diagram. Calves are often bought directly off the ranch or farm by an order buyer. Likewise, calves sometimes will be moved through two or more auction rings, and be owned by several order buyers. (This is one of the greatest inefficiencies in the U.S. cattle marketing system and the cause of considerable cattle sickness and death-loss). The backgrounding yard and the growing yard are sometimes the same operation. Likewise, many feedlots feed high roughage rations as well as high grain rations. In some operations, the cattle are backgrounded, grown, and fed out in the same pen. More often, however, there is at least some separation in these

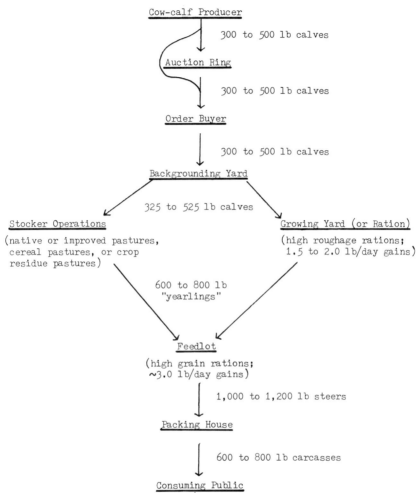

Figure 6-4. STEER CATTLE FLOW CHART (U.S. Cattle Industry)

The flow chart for heifers is similar except that there is a demand in some states (primarily the southern states) for lightweight carcasses. To meet this demand a small percentage of the heifers fed for slaughter are put on high grain rations at what would otherwise be grower weights. Instead of gaining 1.5-2.0 lb/day, they gain 2.25-2.75 lb/day and are therefore fat when they reach 600-700 lb. This is an inefficient process as far as beef production as a whole is concerned. That is, when the feed required for the cow to produce the heifer calf is considered, it is inefficient to slaughter the calf before it can reach its full growth potential. While the process is inefficient for overall production, it is often lucrative for feedlot operators as "light heifers" (calves) typically have 15 to 20% cheaper gain costs than yearling heifers, and yet the fat market for light heifers (600-700 lb) is often only 2 to 5% lower than the market for big heifers (850-1050 lb).

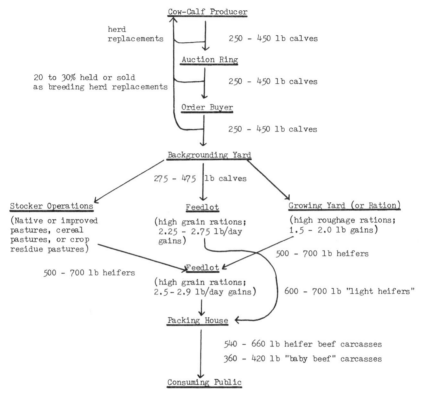

Figure 6-5. HEIFER FLOW CHART (U.S. Cattle Industry)

A portion of the light heifers slaughtered in the South are finished on pasture. They are usually reported as non-fed slaughter or grass-fat cattle, but as a general rule they are fed at least some grain while on pasture, another inefficient process (page 73).

BACKGROUNDING

As shown in the flow chart, most cattle bought as calves are sent to a backgrounding yard before being sent to their final destination. This is because of the high death loss that can occur with shipped in calves. Backgrounding yards are often specialized operations which are geared up to cope with the sickness problems freshly weaned calves usually have. Most traditional-type feedlots do not have the personnel, facilities, or desire to take on the responsibility of backgrounding calves.

Calves are normally held in backgrounding operations about 30 days. The common viral diseases affecting cattle typically have 2 week incubation periods, so if the cattle are healthy after 30 days, the chances of severe disease breaks are greatly reduced (see Chap. 3).

GROWING PROGRAMS

A growing ration (as opposed to a finishing ration) is supposed to allow weight gains without actually fattening cattle. A true growing ration should allow weight gains of about 1.25-1.75 lb/day. Finishing rations typically put on gains of 2.5-3.0 lb/day.

As shown in the cattle flow charts, (Figures 6-4 & 6-5) most cattle (steers) are put on growing rations at weights of 300-500 lb, and are put on finishing rations at weights of 600-800 lb (heifers 500-700 lb). That is, 200-400 lb are put on calves before they are ready to be put on finishing rations. If cattle are put on finishing rations at too light a weight, they become fat too quickly and their performance will be greatly reduced if they are fed to normal slaughter weights.

Growing rations are designed to contain just enough energy to allow the calf to put on muscle tissue (grow) without putting on much fat. Whatever amount of fat that is put on, will reduce performance when the cattle are put on finishing rations. This is because of the phenomenon of compensatory gain (page 140), and because the fat increases the animal's maintenance requirement, and reduces intake. (See also p. 120.)

FINISHING OR FATTENING PROGRAMS

The idea behind finishing programs is to get the cattle to gain as rapidly as possible. To do this, rations as high in energy as possible are used, and they are presented in a manner to obtain maximum intake.

The animals are normally fed until they reach their maximum muscle growth plus a certain amount of fat. Over finishing should be avoided as the laying down of fat is a very inefficient process, which can greatly increase feed costs. This is easy to understand since muscle tissue contains 50-70% water, whereas fat contains very little water. Thus, when an animal is growing and laying down muscle tissue, 50-70% of the weight increase is due to water. For that reason, gain drops off sharply as cattle become fat, but intake decreases only slightly.

Cows are not shown on the flow charts (Figures 6-4 & 6-5), but there are a number of cull range cows fattened in feedlots. Generally they are only fed 45 to 60 days (steers and heifers are normally fed 120-150 days), since being mature they have little or no growth potential; i.e. they have limited potential for putting on muscle tissue . . . most of what they put on is fat.* Occasionally cows are fed longer than 60 days, but it usually results in extraordinarily high gain costs.

THE "CHOICE" QUALITY GRADE MARKETING SYNDROME

Present day marketing conditions (as of 1981) in most of the feeding areas dictate that cattle be fed until they grade low choice by USDA quality grading standards. The main criterion that determines the USDA quality grade is the amount of intramuscular fat or marbling. Some breeds such as Angus have a propensity for marbling, and others such as Charolois do not. Breeds that have a low propensity for marbling must be fed to an excessive amount of finish in order to make the choice grade, since intramuscular fat (marbling) is the last fat to be deposited in an animal.

Consumers on the West Coast do not demand the choice quality grade meat to the extent that consumers in the rest of the U.S. do. As a result, West Coast feedlots are able to market their cattle at a lighter, more efficient weight. Also they are better able to utilize the breeds of cattle that do not have a propensity to marble (Charolois and other large European breeds, Holsteins, Brahma crosses, etc.). Hopefully, consumers throughout the U.S. will learn to accept grades of meat other than choice, and thereby help make the cattle feeding industry more efficient.

THE FEEDLOT CONCEPT OF CATTLE FEEDING

The massing of large numbers of cattle for feeding has come about as a means of justifying large capital expenditures for equipment; in the Midwest for confinement facilities; and in other areas

*Work done at Arizona State Univ. (Wooten et al., 1979) showed cull cows fed for 38 days deposited 75% of their liveweight gain as muscle tissue, and 25% as fat. When fed for 63 days, liveweight gain consisted of 50% muscle, and 50% fat; and when fed for 108 days, 25% liveweight gain consisted of muscle, and 75% was fat. (Data was obtained by slaughtering representative cows before and after feeding and comparing the fed and non-fed carcasses.)

such as the Northwest, for construction of facilities to handle available by-products suitable for feeding (potato waste, etc.).

The technical aspects of cattle feeding remain the same whether the form be the feedlot concept or farmer-feeder. However, the feedlot industry has developed some unique approaches to many of these aspects. For the sake of brevity, the author has combined the discussion of the technical aspects of feeding and feedlot managment practices. In so doing, it is the author's opinion that a more readable text has been the result. The information presented would differ from farmer-feeder operations only in degree and specialization.

FEEDLOT EMPLOYEE STRUCTURE

The employee structure for most of the larger feedlots is very similar. Some yards have unique arrangements, but most have employees that will fall into the following categories.

THE FEEDLOT MANAGER

Figure 6-6. The feedlot manager's primary responsibilities are marketing cattle and buying commodities. Because cattle and grain markets are so volatile, most large feedlots have market teletype terminals installed in the office. Ralph McGuire, McGuire Feedyards, Laverne, Oklahoma.

In the large feedlots the primary functions of the feedlot manager are personnel management, marketing (fat cattle) and commodities buying. Ration formulation and the technical details of feeding are usually left up to the consulting nutritionist. Likewise, animal health programs are usually left up to a veterinarian.

MILL MANAGER OR FOREMAN

Most of the larger feedlots will have at least one man whose sole job is to operate and keep up the feed mill. Of necessity, most mill managers are reasonably knowledgeable mechanics and electricians, as machinery breakdowns are inherrent problems with feed mills.

It is the responsibility of the mill manager to see that the feed is processed correctly, and with some grains and processes, this is of vital importance. With steam-flaked milo, for example, if the grain is not steamed long enough, or if the rolls are not set under enough pressure, cattle performance could be reduced as much as 8%. A 10,000 head feedlot will process about 170,000 lb of grain a day; at 6¢/lb an 8% inefficiency would cost the feedlot (owners of the cattle) about $816/day. Obviously the feed mill manager must be a conscientious person.

Figure 6-7. Feedlot mills are complex machinery systems. As a result, feedmill foremen must be reasonably capable electricians and mechanics, as well as knowledgeable in feed processing.

BUNK READER OR FEED CALLER

The job of the bunk reader or feed caller is to estimate how much feed a pen of cattle will consume for a given period. He then turns his estimates in to the feed mill or to the feed truck driver directly, and that is how much ration each pen gets.

In feedlots of over 20,000 head, bunk reading is usually a full time job and it is always a very important job. In order to keep intake high, cattle must always have fresh feed in front of them. If feed becomes stale or moldy, consumption and performance drops off significantly. With processed rations most feedlots try to feed at least twice a day, and want the cattle to have just cleaned up the previous feeding, before the next feeding. The bunk reader must therefore be skillful and experienced, taking into account the type of ration, the sex, weight, breed, and condition of the cattle, as well as the weather.

FEED TRUCK DRIVERS

The feed truck drivers obviously drive the feed trucks and deliver feed to the bunks. The ration for several pens of cattle are usually put on one truck and only the amount of feed called for by the bunk reader is given. Most of the feed trucks have electronic scales and the required amount of ration is metered into the feedbunk. By experience the truck driver knows how fast to drive in order to auger off the required amount of feed.

The turnover rate in feed truck drivers is usually quite high. Having inexperienced or otherwise poor quality employees driving feed trucks is an unwise, but all too often situation. One of the biggest problems that occurs is when the driver feeds the wrong pen . . . the worst situation being when a pen of new cattle receive the "top" ration, thereby causing acidosis and possible death loss.

Figure 6-8. Feedlot cowboys are charged with the responsibility of locating sick animals. Riding through the same pens of cattle all day - every day is extremely monotonous, yet requires a very observant individual. With feedlot cattle worth $300-$800/hd, and a ratio of 2,000-5,000 hd/ man, the necessity of a quality employee is obvious.

COWBOYS

The primary job of the feedlot cowboy is to locate and cut out sick cattle. In some yards the cowboys doctor the cattle they cut out, but in most yards they turn them over to a seperate doctoring crew. In the larger, more progressive feedlots the cowboys fill out a card on each animal they pull; identifying the pen, animal's tag number, the condition the animal was found in, and the suspected disease or malady.

DOCTORING CREW

As the name implies, the doctoring crew's primary function is to treat and care for the sick cattle brought to them by the cowboys. There will usually be a "head doctor", and quite often these are men relatively well versed (for laymen) in veterinary medicine. Usually they receive programs and guidance from consulting veterinarians. In the larger feedlots these men obtain a lot of experience and often become remarkably adept at diagnosing and treating a wide variety of diseases.

Usually these men take a lot of pride in their work, and some even consider themselves professionals. This is no doubt because

Figure 6-9. Feedlot "doctor" conducting autopsy. Although few feedlot "doctors" have formal training in veterinary medicine, many become reasonably proficient in diagnosing and treating diseases. This comes by association with consulting veterinarians and constant daily contact with diseased animals.

they come in contact with a wide variety of diseases and treatments, and thus do not get bogged down in the monotony that typifies most feedlot work.

PROCESSING CREW

Most feedlots have a separate processing crew which dip, brand, castrate, vaccinate, implant, etc., incoming cattle. In some cases the processing crew may not actually be employees of the feedlot, but rather a contract crew working part time. Processing crews usually consist of one experienced man with some technical knowledge and several younger inexperienced men. It is important that the experienced man be there as processing crews handle vaccines, medications, and growth hormones.

PERSONNEL MANAGEMENT

As with most businesses, one of the biggest problems faced by large feedlots is labor. Feedlot work is dirty, monotonous work and so the turnover in employees is usually quite high. But while the work appears very routine and is monotonous, it requires a very observant individual. The difference between good and poor performance of cattle in a feedlot depends upon a lot of subtle indicators, and unless the feedlot employees are willing to be observant and take initiative, overall feedlot performance suffers accordingly.

The two most common mistakes in personnel management made by feedlot managers are working their men too hard, and assuming that they will accept responsibility. The amount of cattle in the feedlot can fluctuate significantly, and when large numbers of cattle come in at once, there is a tendency for many feedlot managers to demand too much of their employees (this is the most common personnel management problem in all Amercan agriculture). It doesn't take many 12 to 16 hour days before the employee loses zest for his job. When this happens, many sick cattle don't get noticed (unless acutely ill), the doctoring crews will spend less time trying to diagnose the disease and will simply "give 'em a shot of something", feed truck drivers will feed over stale or moldy feed left in the bunk, mill men will not be as conscientious in processing the feed, etc., etc.

As mentioned earlier, the second biggest mistake made by many feedlot managers is to expect that their employees will accept responsibility; i.e. take the initiative to do something when they can see that it needs to be done. Most employees require supervision, not oppressive or critical supervision, but a regular unoffensive, discreet supervision to see that things are being done properly. The manager must realize that most minimum wage hourly workers do not have the basic knowledge or understanding of what needs to be done, as well as possible lack of initiative. A common problem is for feedlot managers to become preoccupied with office work and fail to get out and actually see what is going on in the yard.

Feedlot labor is often hard to come by, but the feedlot manager should strive to get as good a quality a man as possible. The manager must remember that to get a good man, he's going to have to pay more than the going minimum wage. Often it doesn't take more than $200-$300/month extra to get a good man, and a good man can easily save that much in reduced machinery and vehicle damage alone.

Figure 6-10. A good feedlot employee will usually cost more in terms of wages, but a good man can often save the extra expense just in terms of reduced equipment and vehicle damage.

BASIC FEEDLOT NUTRITION

Nearly all of the larger feedlots leave nutrition and ration formulation up to a consulting nutritionist. Many feedlot managers have some technical background in nutrition, but the use of additives, NPN compounds, and their interaction with different feedstuffs has become so complex, that most feedlots rely on professional nutritionists. It is important, however, that feedlot personnel have a basic understanding of feedlot nutrition and ration formulation. The following sections are designed to give the reader a basic understanding of feedlot nutrition and the principles involved in formulating rations.

GRAIN PROCESSING

The subject of grain processing is a very complex and often confusing subject. The question most often asked of nutritionists is "What is the best method of processing?". The answer is that there is no one best method. Grains react differently to different processes, and different processes will have advantages and disadvantages with different associative factors.

DRY-ROLLING AND GRINDING

Dry-rolling is usually accomplished via a roller mill which consists of two metal rolls spaced fairly close together. Grain is passed between the rolls and is thereby cracked or rolled. Grinding is usually accomplished via a hammermill or other type of machinery that pulverizes the grain. The product obtained from grinding is usually much finer and more floury.

With every grain except milo, rolling is usually preferable to grinding. If done correctly, rolling will break or crack corn or wheat into several large pieces which is preferable to the large amount of fines produced by grinding.

Animal performance on dry-rolled corn, wheat, and barley is usually better than when those two grains are ground. The author theorizes that by having large pieces of grain (rather than flour), more of the grain bypasses the rumen and is digested enzymatically in the lower tract, thereby eliminating some of the inefficiency of rumen digestion (page 190). In addition, consumption is usually higher on dry-rolled grain than finely ground grain.

With milo, very fine grinding is preferable to coarse rolling. For best performance, milo should be heat-treated (steam flaking or popping) or be used as a high-moisture or reconstituted grain. If dry-rolling or grinding is the only process available, the milo should be ground as fine as possible.

The chemistry of milo is not completely understood, but apparently the matrix surrounding the starch and protein is much more difficult to digest than other grains. The mechanical breakdown of the matrix (via fine grinding) apparently enhances digestion as cattle usually perform better on finely ground milo than coarsely ground or rolled milo. (Whenever milo is finely ground, ration conditioners such as molasses or silage should be used.)

STEAM FLAKING

In steam-flaking, grain is exposed to live steam for 15 to 30 minutes, and then run through a large pair of rollers held to a close tolerance.

Steam-flaking is distinguished from steam rolling, by the steaming time and the amount of pressure placed on the grain when run through the flaking rolls. An ordinary roller mill is incapable of exerting enough force to truly flake grain. As illustrated in Figure 6-11 & 6-12, a flaking mill consists of rolls that are much larger and heavier than those used in ordinary roller mills. In addition, hydraulic rams or very heavy springs are used to maintain pressure on the grain passed between the rolls.

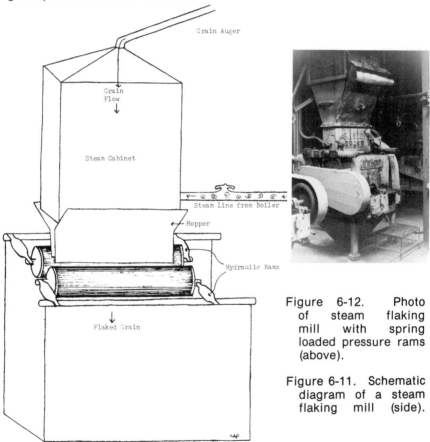

Figure 6-12. Photo of steam flaking mill with spring loaded pressure rams (above).

Figure 6-11. Schematic diagram of a steam flaking mill (side).

Simply rolling grain after steam treatment will not produce the performance response of flaking. The tremendous pressure exerted upon the grain during flaking breaks down the starch and protein matrix and makes the grain more digestible.

The response to steam-flaking varies with the grain used. Milo shows the greatest improvement by far. Steam-flaked milo can be fully 10% more digestible than dry-rolled or ground milo. In addition, consumption of flaked milo is often greater than dry processed milo, and so animal performance often shows improvement in excess of 10%.

Corn does not show the response to steam-flaking that milo does. Generally about a 2 to 4% response (over dry-rolling) is all that can be expected.

Many of the large feedlots utilizing steam flaking have switched from milo to corn (recent large scale irrigation has made corn a major commodity in the Southwest). The reason for the switch has been consistency of performance. Many of the large feedlots have found their performance with steam-flaked corn seems to be much more consistent and predictable. The reason is usually attributed to the variability in the different varieties of milo.

Certainly it is true that milo is a much more variable commodity than corn. However, it is the author's opinion that the principal reason for poor performance with steam flaked milo (as compared to corn) has been due to imperfect quality control during processing, rather than varietal differences per se. In order to get the full value out of milo with the steam-flaking process, quality control must be exacting. The grain must be left in the steam cabinet long enough for moisture to penetrate the entire kernel (raise kernel moisture 7-9%), and the pressure exerted on the grain when run through the rollers must be great enough to produce a true flake. Without a thin, true flake, performance with milo will drop off sharply.

With corn, performance is not that much better with a true flake. Hence, imperfect mill management doesn't reduce performance nearly so much. It is for this reason that the author feels many feedlots have had much more success steam flaking corn than milo.

Wheat and barley apparently respond very little to steam flaking. Consumption is usually a little better, and so feedlots that have steam processing equipment will often go ahead and steam process wheat and barley. Typically, time in the steam cabinet is reduced to 5 or 10 minutes, and pressure on the rolls is also reduced. Reduced pressure with wheat is usually a necessity, as the gluten will often gum up the rolls if an attempt is made to flake it.

PRESSURE FLAKING

Pressure-flaking is very similar to steam-flaking, except that a pressurized steam chamber is used. By placing the grain under pressure with steam, time in the steam chamber can be greatly reduced. Due to the expense and danger involved with pressure flaking, it is not a popular processing method.

POPPING

Popping is a processing method developed several years after steam-flaking. In true popping the grain is heated until its internal moisture causes the kernel to rupture. In the processing of grain for feedlot use, the grain is usually heated to 450 -550° F to produce about a 30% pop. The grain is then run through a roller mill, and when the pressure from the rolls is exerted on the grain, the unpopped kernels usually go ahead and pop. The reason only an initial 30% pop is strived for is simply to save energy. Rolling is required to reduce the bulk density of the feed. Without rolling, a popped ration would be too fluffy and would reduce intake. Popping reduces moisture in the grain, and so 3-6% water is usually added back to the grain after processing.

With popping, milo is again the grain that responds the most. When done correctly, popped milo will perform as well as steam-flaked milo. As with steam-flaking, close attention must be paid to the processing.

Popping is much more energy efficient than steam-flaking. This is because dry heat is applied directly to the grain, rather than in the form of steam. Water requires over twice as much energy to change its temperature, and thus the creation of steam uses more energy than applying the heat directly to the grain.

The main disadvantage to poppers is that the grain must be very clean, or dust explosions will occur. Usually these explosions are more disconcerting than destructive, but they do pose a danger.

HIGH-MOISTURE GRAIN (early harvested grain)

High-moisture grains are being used extensively in the cattle feeding areas, and to a certain extent are a somewhat controversial issue. High-moisture grain is defined as grain harvested at about 22-30% moisture and either ground and ensiled, treated with organic acid preparations as a preservative, or stored in oxygen limiting structures.

The feeding value of high-moisture milo is substantially better than dry-processed milo. Research data varies, but apparently high-moisture milo approaches the value of steam-flaked or popped milo (on a dry matter basis).

High-moisture corn apparently has about the same energy value as dried corn on a dry matter basis. It is difficult to compare high-moisture grain to dry grain because high-moisture grain

Figure 6-13. High-moisture milo being ground and put in a bunker silo for ensiling. (Felt Feedyard, Felt, Okla.)

Figure 6-14. Large bunker of ensiled high-moisture corn. (Master Feeders, Hooker, Okla.)

has à number of feeding idiosyncrasies. To begin with, high-moisture grain should not make up the entire grain portion of a finishing ration. At least 20-30% dry grain should be added. Without additional dry grain, consumption tends to decline. Some dry roughage should be added to a high-moisture grain ration. Silage may be used, but only sparingly. Ground or chopped alfalfa hay is by far the best roughage for high-moisture grain rations.

High-moisture grain is relatively high in NPN and soluble nitrogen. The amount will depend on the moisture; the higher the moisture (the more immature the grain is), the higher the level of NPN will be. Apparently the protein content of the grain is not fully formed until the grain becomes mature (field dry). Being high in NPN, high-moisture grain protein supplements must be lower in urea than supplements designed for use with dry grain. If normal levels of urea are used, it will usually result in reduced cattle performance.

Cattle fed high-moisture grain rations must be fed at least twice a day (preferably three times a day), and must be allowed to clean up between feedings. High-moisture grain molds rapidly, and this will reduce consumption and performance if left in the bunk.

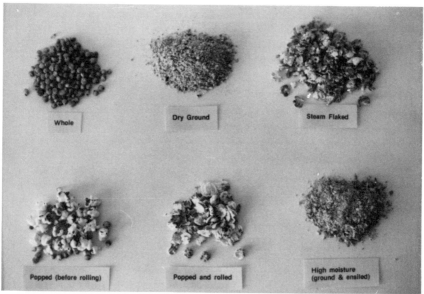

Figure 6-15. Evolution of processing methods with grain sorghum (milo), the crop that was responsible for the establishment of the south-western feedlot industry. With the development of hybrid varieties that could be grown in an arid climate, the Southwest emerged as a cattle feeding area. The development of steam flaking meant that a feedlot that could afford the equipment could substantially outperform the feedlot with dry rolling or grinding equipment. As a result, commercial feedlots were forced to expand their capacities to justify the expenditure of steam processing equipment . . . and the era of the large feedlots began. Popping, another capital intensive form of processing came later. At the current time, due to skyrock-eting energy costs, the use of high moisture milo is beginning to receive attention.

RECONSTITUTED GRAIN

Reconstitution means adding water back to dry grain to bring the total moisture up to 24-30%. The only grain that responds significantly to reconstitution is milo. When done correctly, reconstituted milo, like high-moisture milo, approaches the value of steam-flaked or popped milo.

To reconstitute milo it must be stored whole in some sort of oxygen limiting silo. The usual treatment is to add water and store the grain for at least 14 days. The grain is removed and ground or rolled just before feeding. Grinding dry milo, adding water, and ensiling in a pit (as commonly done with high-moisture milo), does not produce the response obtained when milo is stored whole and then ground.

TEMPERING

The practice of tempering (soaking) grain in water before feeding has been carried out for many years. Soaking may increase performance (over dry grains) due to increased palatability and intake, but it does not measurably improve digestibility.

WHOLE CORN FEEDING

The feeding of whole shell corn has become fairly popular in the last few years. In the cattle feeding areas, whole corn feeding is a controversial topic as some feeders have had excellent results and others have had relatively poor results. The controversy stems from the fact that whole corn feeding has a somewhat different digestion pattern than processed grain, and requires a much different type of management. The feeders that have recognized this have had good results, and those that have tried to feed whole corn as they would processed grain, have had poor results.

With whole shell corn the cattle chew the corn and break it into several large pieces. The surface area for bacterial digestion in the rumen is therefore not nearly as great as with processed grain. As a result, much of the corn escapes rumen fermentation and is passed to the lower tract. Thus, much of the inefficiency of rumen fermentation is by-passed.

Since much of the digestion takes place in the lower tract, a protein supplement of high biological value is required. The use of typical high urea supplements with incomplete natural protein bases will produce unsatisfactory results.

Since fermentation is decreased, finishing rations with whole shell corn can be used with little or no roughage. Roughage is not utilized in a finishing ration very well (due to low concentrations of cellulolytic bacteria), and so this is a distinct advantage for whole corn feeding.

Figure 6-16. Feedlot cattle receiving a ration of whole shell corn without roughage (Figure 1 Feedlot, Booker, Texas.)

Cattle fed whole shell corn should never be allowed to clean up (run out of feed), or they will become hungry and tend to swallow their corn on the next feeding. Keeping feed in the bunk all the time does not cause the mold or stale feed problems normally encountered with processed rations since the corn is dry.

Done properly, whole corn feeding will produce at least as good as, and usually better, results than processed corn. Two to three percent of the corn will pass through the cattle whole, but the reduced losses from rumen fermentation usually more than offset the unchewed grain loss.

USE OF BY-PRODUCT FEEDS

There are a number of by-product feeds that can and have been used for all kinds of cattle. Examples would be cull fruits and vegetables, cannery wastes, citrus pulp, peanut or pecan hulls, etc., etc. Some of these products have a relatively high feed value, and others extremely poor. The products that are of high value are usually well known and often bring a price equal to that of a conventional feed with similar food value; e.g. citrus pulp.

Before utilizing one of these materials, professional nutritional advice should be sought. These products are so different and their actual use would depend on so many extraneous factors that specific recommendations simply are not feasible in the confines of this text. Probably the only generality that can be made is that individual products are usually quite variable. Products such as peanut hulls can vary from 3-4% CP to as high as 12-14% CP, depending upon the number of immature nut meats present. In other products the moisture can vary radically. Chemical analyses often show large variations in ash, as well as individual minerals, which is usually a reflection of the amount of dirt present. Some products, such as cull onions, may contain alkaloids or other compounds which, if fed in large amounts, can cause problems. Again, professional advice should be sought before embarking on a by-product feeding program.

In addition, the feeder should be certain that the by-product will be available in sufficient supply to justify putting it in the ration. Such products are often seasonal, and sometimes delivery even within the season can be sporadic. Rumen microorganisms take a period of time to adapt to feedstuffs, and cattle performance can therefore be hindered by having to change rations to conform to commodity availabilities. Consequently if the feeder cannot rely on a daily supply of the by-product throughout the feeding period, it would be prudent to reconsider using it.

FEED ADDITIVES

DES (Diethylstylbesterol)

DES was the first major feed additive developed and widely used by the cattle feeding industry. A synthetic estrogen (female sex hormone), DES increases nitrogen retention, resulting in a higher rate of gain. Vast amounts of research have shown DES to increase daily gains up to 10% when used as either an oral feed additive or as an implant placed under the skin (of the ear).

DES is one of the few compounds that will actually change carcass composition. By increasing protein deposition, DES usually reduces intramuscular fat (marbling) and thus results in a lower carcass grade.

Reported to cause cancer in laboratory animals, DES was

banned by the Federal Drug Administration (FDA) in 1973. The ban was later rescinded by a court order resulting from a law suit challenging the FDA's action. In 1979 DES was banned again, and was still banned at the time of writing.

MGA (Melengesterol Acetate)

MGA is a synthetic progesterone which when fed to feedlot heifers results in improved gains and feed conversions. Progesterone is the female hormone which supresses the heat cycle, and thus by feeding a synthetic progesterone, the heat cycle of feedlot heifers is supressed. The estrous cycle recurs about every 21 days, and so without MGA heifers go off feed for 1 or 2 days each month and cause disturbances in the pen. With MGA, the heifers usually do not come into heat, and thus stay on feed and gain better.

RUMENSIN (Monensin Sodium)

Rumensin, the trade name for monensin sodium, is a antibiotic originally used as a coccidiostat in poultry rations. Only recently it was discovered that the product reduces methane formation in ruminants. Because of this property, it has become a very useful cattle feed additive.

Methane (CH_4), a product of rumen fermentation, is an energy loss (page 190). By reducing the amount of CH_4 given off, an energy saving results. This increased the amount of propionic acid in the rumen. Propionic acid is used more efficiently for weight gain than other major acids (acetic, butyric).

High grain rations produce more propionic acid than roughage rations. Propionic acid in the bloodstream controls the appetite response in cattle. When Rumensin is used in conjunction with high grain rations, the blood propionic level is increased to the point where intake is reduced. As a result, the cattle will gain at the same rate on a reduced amount of feed. (If excessive amounts are fed, however, cattle can go off feed.) On high roughage rations the intake reduction response does not ordinarily occur, and so gain response is experienced.

As added benefits the methane reduction effect significantly reduces bloat, and on high grain rations the reduced intake effect has substantially reduced acidosis and founder problems. The product is also apparently useful as a coccidiostat in cattle, since clinical cases of coccidiosis appear to be reduced in feedlot cattle when it is used as an additive.

ANTIBIOTICS

Numerous antibiotics are used as feed additives either as a low level growth promotant or at a high level as a means of controlling subclinical bacterial infections. Low level feeding usually means 60 to 100 mg/hd/day and high level from .7-1.0 g/hd/day.

IMPLANTING

The term implanting refers to the use of growth stimulating compounds in the form of a pellet, deposited under the skin of the ear. The reason the ear is chosen as the site, is because it is removed from the carcass at slaughter, so there is no opportunity for unabsorbed residue to contaminate the meat.

At the time of writing there were two commercial products available for use. (1.) Ralgro, the trade name for Xeranol, marketed by IMC, is cleared for use in baby calves, and; (2.) Synovex, marketed by Syntex, is a sex specific product containing naturally occurring hormones cleared for calves 400 lb and larger. Designated as Synovex H and Synovex S, "H" is for heifers and "S" for steers. In years past, DES was the standard of the industry, but as mentioned earlier, has been outlawed by the FDA.

The use of implants will normally increase gain from about 6-12%, and have therefore become a standard practice in the feedlot industry. When used in feedlot cattle, one must be careful to observe the withdrawal times set by the FDA. Cattle implanted with Ralgro may not be slaughtered for 65 days and Synovex for 60 days (DES had a 120 day withdrawal). For "long fed" cattle (160 days or more), there is university research showing a response to a second implant given near the middle of the feeding period (Wagner, 1976; and Matsushima, 1975). Table 6-1 represents Matsushima's data:

Table 6-1. RESPONSE TO IMPLANTATION AT VARYING TIMES.

	Control (no implant)	Synovex S at Day 1	Synovex S at Day 48	Synovex S at Day 1 & Day 48
No. of steers	60	60	60	60
Initial wt., lb	710	710	710	710
Final wt., lb	1091	1150	1170	1162
Ave. daily gain (160 days)	2.38	2.75	2.88	2.83
Feed conversion	9.88	9.29	8.67	9.09

Note that the cattle receiving only one implant on day 48 outperformed all the other treatments. Matsushima reported less bulling in those steers, but did not speculate as to why they were more efficient.

During the physical act of implanting, care should be exercised to be sure that; (1.) pellets are placed in the correct portion of the ear, and; (2.) pellets are not crushed during the implantation procedure. When using the Synovex product, a common mistake has been to place the pellets too close to the base of the ear. The old DES implant was designed to go ⅓ of the length of the ear from the base, and Ralgro is designed to be placed within 1 inch of the base of the ear. Synovex is designed to be placed in the middle of the ear (see Figure 6-17).

Placing Synovex too close to the base of the ear will result in too rapid absorption of the hormones, which can reduce performance and conceivably could increase bulling activity. Also, after the implanting needle has been driven into the ear, pull it back ⅜ to ½ inch before ejecting the pellets. This will leave a space for the pellets to go and reduce the chances of pellets being crushed. Crushed pellets will also lead to excessively rapid absorption of the drug. As a final note, implanting needles should be kept clean as a dirty needle can lead to infection. If infection occurs, scar tissue will form around the pellets and reduce or eliminate assimilation of the drug.

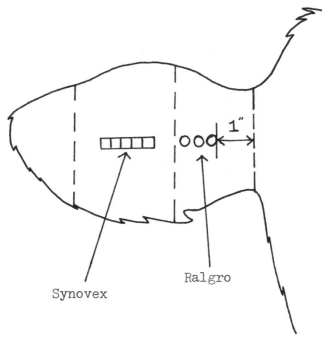

Figure 6-17. Proper site of implantation.

CATTLE MALADIES PECULIAR TO FEEDLOTS

ACIDOSIS AND FOUNDER
Described in detail on page 121, acidosis is a fairly common problem. High energy rations lower the pH of the rumen and if cattle are put on too high an energy ration too quickly, the pH will drop to the point of causing distress and sometimes death.

FEEDLOT BLOAT
Bloat is of course not restricted to the feedlot, but the type of bloat encountered in the feedlot is usually different than typical pasture bloat. As explained on page 78, pasture bloat is caused by accumulation of foam in the rumen which inhibits the eructation response. Feedlot bloat is generally characterized by gas with much less foam. Some feedstuffs and combinations of feedstuffs are known to increase the incidence of feedlot bloat. Of the grains, barley has the reputation of causing the most bloat, particularly when fed in combination with alfalfa as the roughage.

Figure 6-18. Holstein steer with feedlot bloat. Notice the distended rumen on the left side, just behind the ribs.

Treatment for feedlot bloat is similar to pasture bloat with the exception that sometimes passing a tube down the esophagus into the rumen will relieve gas pressure. With pasture bloat that treatment is usually not effective since a greater accumulation of froth blocks the air passage in the tube.

Quite often feedlot animals that bloat become chronic bloaters. There is some research to indicate that this may be a genetic fault, but whatever the reason these animals become a nuisance. Feedlots often eliminate the problem by simply hauling them off to a salebarn. That certainly doesn't solve the problem for the beef industry as a whole, but does eliminate the problem for the individual feedyard.

Since the introduction of Rumensin the incidences of feedlot bloat have been greatly decreased. As explained previously, Rumensin reduces fermentation in the rumen which therefore reduces gas formation. Before Rumensin, fat was often added to reduce bloat as it also has an inhibitory effect upon gas production.

URINARY CALCULI (Waterbelly)

Urinary calculi is the development of burr shaped stones made up of dietary minerals within the kidneys or bladder of steer cattle. When the stones attempt to pass from the bladder they often become lodged in the urethra. This makes urinating difficult or impossible. Left untreated the bladder will often rupture and the animal dies of uremia. If the stones become lodged near the prepuce, sometimes the penis itself will rupture causing an absolutely horrible infection.

The problem usually only occurs in steers, since the size of the urethra is reduced due to castration. Stones can form in heifers and bulls, but the urethra is usually large enough to allow the stones to pass.

In many cases surgery is the only cure. Most often the stones accumulate in the S curve of the penis known as the sigmoidal flexure. The veterinarian will make an incision below the anus, sever the penis, and bring it out through the incision. The steer then urinates similar to a heifer.

There are two types of urinary calculi, pasture and feedlot. The difference is in the mineral which makes up the stone. Stones formed on pasture are usually a silica base. Stones formed in the feedlot are made up of calcium, magnesium, and phosphorous.

Of the clinical cases occurring in the feedlot, rations consisting of milo are most often implicated. Urinary calculi on other grains are much less common. Specifically, urinary calculi appear most frequently when the Ca:P ratio in the ration comes close to 1:1. If care is not taken to ensure at least a 1.5:1 Ca:P ratio, as many as 8% of the cattle can come down with clinical cases of calculi.

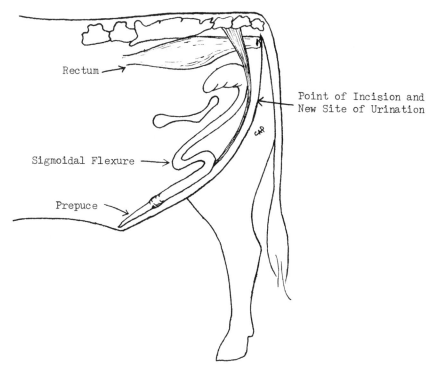

Figure 6-19.   SURGICAL CORRECTION OF URINARY CALCULI IN STEERS.

Normally 50-60 days are required for feedlot calculi to develop. If clinical cases appear before this time, it is usually due to silica stones formed on pasture before the cattle were brought in.

Prevention, as mentioned earlier, consists of ensuring a 1.5-2.0:1 Ca:P ration when using milo grain. Even so, some feedlots still have problems with urinary calculi. In feedlots with a history of the problem, the addition of ammonium chloride or ammonium sulphate in the ration will do much to eleviate the problem. These compounds acidify the urine slightly, which apparently prevents the precipitation of Ca, P, and Mg and subsequent stone formation. Levels commonly used vary from 14 to 28 g/hd/day. (These compounds are not effective against silica or pasture formed stones.) In addition, some feedlots will feed extra salt in the ration to increase urine flow.

THE BULLER SYNDROME

For a yet unexplained reason, cattle of the same sex (steers and heifers), will single out individuals within the herd and mount them as in breeding. Why only individuals are singled out for this seemingly homosexual behavior is not known. It has been theorized that the "Buller" animal gives off a pheromone (hormone like odor), but that has not been proven. By observation, it often appears that it is the smallest and weakest animal that is singled out.

The situation is often serious as many times the buller animal is ridden until it is too weak to make it to feed and water. Unless cut out of the pen, buller animals can die of dehydration. At the very least, their gain performance will be very poor.

The buller syndrome is not totally restricted to feedlot cattle, as it does occasionally occur in pasture cattle. But because of the crowding due to confinement, and the more common use of hormonal growth implants in feedlots, the problem is much more serious. The only cure is to pull the buller animals out of their pens and place them in a separate pen. Nearly all feedlots have one small buller pen for this purpose.

The problem is pretty much inherent, and it can be expected that approximately 2-5 buller animals will have to be cut out for every 1,000 head of cattle. During spring and early summer the problem usually reaches its greatest intensity.

Occasionally feedlots experience extreme outbreaks of bulling. Usually this situation can be traced to faulty implant technique; either crushing the implant pellets by not backing the needle out before driving them into the ear, or placing the pellets too close to the base of the ear.

POLIOENCEPHALOMALACIA

The term polioencephalomalacia means a cerebral (brain) disorder which results in a lack of motor control of the nerves. There are a number of disease-causing organisms that can create a form of encephalomalacia, among them are Haemophulos somnus bacteria, the nervous form of coccidiosis, and certain mycotoxins found in molds.

The term polioencephalomalacia is most often used in reference to an encephalitic condition in feedlot cattle caused by an apparent thiamin (vitamin B_1) deficiency. For some unknown reason feedlot cattle will sometimes develop a thiaminase (thiamin-destroying enzyme) in the rumen. Thus, thiamin deficiency symptoms result, even though there is what would otherwise be an adequate level of thiamin in the diet. The cattle become spasmodic and will generally have the head bent backwards; with more advanced cases, they will be blind and will have convulsions. Treatment consists of injection with thiamin and dextrose. If left untreated, the condition is usually fatal.

PROLAPSE IN HEIFERS

A relatively common problem in feedlot heifers is vaginal prolapse. Some feedlots have reported incidences as high as 1%. Occasionally the anus will also become prolapsed.

As can be seen in Figure 6-20, the vaginal and/or anal wall is forced out of the body. In severe cases even the uterus will become exposed. Treatment consists of forcing the tissue back into the body cavity and suturing it in place.

The cause of prolapse is not clear. It is thought that the feedlot use of implanted steroids is a contributing factor. It is also thought that the use of MGA is somewhat protective. Although plausible, neither of these ideas have been accurately documented.

Figure 6-20. Heifer with vaginal and anal prolapse.

FEEDING MANAGEMENT

BUNK READING

With processed grain rations, it is vitally important to have fresh feed in front of the cattle at all times. Feed left in bunks over 12 hr will usually become stale and begin to mold. Stale feed, of course, reduces intake and subsequent performance.

For every 1% reduction in intake, reduced gains of 1.5-2.0% can be expected.

In order to make sure the ration in the bunk is always fresh, most feedlots feed at least twice a day. With high-moisture rations, some yards feed three times a day. Between feedings, cattle should be allowed to "clean up" (finish all the ration from the previous feeding). Ideally, cattle should be fed no later than ½ hour after they have cleaned up. In order to keep this kind of schedule, most feedlots will have at least one man assigned to reading bunks.

It is the job of the bunk reader to determine how much feed each pen of cattle will consume for the given period of time. In order to do this, a number of factors must be taken into consideration.

CONDITION OF THE CATTLE
Newly arrived cattle are often very difficult to get started eating, particularly cattle that have never been penned up before. Once they begin eating, consumption will usually increase until they begin to fatten. As cattle become fat, their intake gradually drops off.

AGE OF THE CATTLE
Calves and yearlings tend to eat at a level in relation to their weight, until they start to become fat. Old mature animals such as cull cows and range bulls often eat much more in relation to their weight than younger animals.

BREED OF THE CATTLE
There is little difference in consumption for most of the beef breeds. The larger breeds such as Charolois will eat more than smaller breeds, but the increase is usually in relation to their size. However, large dairy breeds (Holstein and Brown Swiss) will eat a disproportionately larger amount of feed for their size than most beef breeds.

SEX OF THE CATTLE
As a general rule, steers will eat 5 to 10% more than heifers of the same weight. However, mature cows will nearly always out consume yearling steers of the same weight.

TYPE OF RATION
There are three main factors in feedlot rations that affect intake; moisture, amount of fiber, and energy level.

If energy and fiber are held constant, cattle will usually eat

Table 6-2. APPROXIMATE INTAKES ON VARIOUS RATIONS*

Type of Cattle	1 65% steam flaked grain 35% dry roughage	2 90% steam flaked grain 10% dry roughage	3 50% dry processed grain 50% silage	4 75% dry processed grain 25% silage	5 all silage ration	6 100% concentrate whole corn ration
400 lb steer calves	8-14 lb	NA	15-24 lb	NA	28-38 lb	NA
750 lb yearling steers	25-32 lb	23-26 lb	32-38 lb	25-30 lb	42-54 lb	21-25 lb
750 lb yearling heifers	23-29 lb	19-24 lb	28-35 lb	23-27 lb	38-47 lb	17-22 lb
1,000 lb steers	24-30 lb	21-25 lb	28-36 lb	24-28 lb	39-47 lb	18-23 lb
1,000 lb Holstein steers	27-35 lb	24-28 lb	36-45 lb	27-34 lb	48-60 lb	23-27 lb
900 lb cull beef cows	26-36 lb	22-28 lb	36-46 lb	25-34 lb	48-70 lb	NA

*Rations are assumed to contain proper protein/mineral supplementation. Cattle assumed to be adjusted to feedlot conditions. For rations w/Rumensin (30g/ton) reduce consumptions 6-10% except for ration #5.

NA - This type of ration not usually fed to that class of animal.

more pounds of high-moisture rations to a proportion of the higher moisture content. With extremely moist feeds, the moisture itself may limit intake.

Fiber tends to limit intake since it takes more time to digest than other nutrients. Thus it has a tendency to accumulate in the rumen and make the animal feel full. Therefore, high roughage rations tend to limit intake.

High energy rations tend to limit intake through a feedback mechanism that shuts off appetite when the energy content of the blood reaches a certain level. Simple stomached animals use blood glucose (a simple sugar) as the appetite depressor (this is why humans lose their appetite after eating sweets). Ruminants use propionic acid. Propionic acid is produced in greater quantities (in the rumen) from rations high in concentrates. Cattle therefore usually consume feed to a biologically preset energy or concentrate level. Given rations with grain levels from about 65% to 90%, cattle will eat about the same amount of grain. Thus, if the ration has 35% roughage, cattle will eat more pounds of that ration than the ration containing only 10% roughage. (Since roughage is poorly utilized in high concentrate rations, as low a roughage ration as possible is always the most efficient ration.) At a point somewhere over about 35% roughage (depending on the type of roughage), the increased fiber content will begin to slow down intake.

RUMENSIN

Rumensin decreases intake of high concentrate rations significantly. Rumensin decreases methane loss, and so performance is not hindered by the reduced intake. Depending upon the level used, high concentrate ration consumption will be reduced by 5-10%. Consumption on high roughage rations is not reduced.

WEATHER CONDITIONS

During extreme temperatures, hot and cold, consumption usually drops off. During moderately cold weather consumption typically increases. (Relative terms such as moderate and extreme are used since the temperature cattle are adjusted to will greatly affect how they react to varying temperatures. Cattle accustomed to 0°F weather can be heat stressed at 60°F, and cattle accustomed to 110°F temperatures can be chilled by 60°F weather.)

Cattle are very sensitive to weather changes, and will increase their consumption as much as 20% just before storms. Whenever the barometer begins dropping or inclement weather is otherwise predicted, it is a good idea to increase the amount of ration to be fed.

THE IMPORTANCE OF INTAKE

Intake is extremely important because the greater consumption is, the greater the daily weight gain will be, which in effect lowers the feed conversion and cost of gain. This is because as the cattle eat more, their maintenance requirement becomes a smaller percentage of total consumption. For example, a 992 lb steer has a maintenance requirement for 7.9 lb of TDN. If a ration is fed that is 78% TDN, then the steer must eat 10.1 lb of the ration just to meet his maintenance requirement (7.9 lb ÷ .78= 10.1 lb). Everything he eats over 10.1 lb will go toward weight gain. Using data from NRC (Appendix II), look at how performance and cost of gain varies with intake (ration cost @ 6¢/lb):

Table 6-3. COST OF GAIN AS AFFECTED BY INTAKE.

Intake Ration*	20 lb	21 lb	22 lb	23 lb	24 lb
lb TDN	15.6	16.4	17.2	17.9	18.7
Less lb TDN for maintenance	-10.1	-10.1	-10.1	-10.1	-10.1
lb TDN left for gain	5.5	6.3	7.1	7.8	8.6
Daily gain	1.8 lb	2.25 lb	2.5 lb	2.95 lb	3.15 lb
Conversion**	11.1:1	9.3:1	8.1:1	7.8:1	7.6:1
Cost of Gain***	66.6¢/lb	56¢/lb	52.8¢/lb	46.8¢/lb	45.7¢/lb

*78% TDN ration
**Intake ÷ gain
*** @ 6¢/lb ration cost x conversion

PEN AND BUNK SPACE REQUIREMENTS

Pen and bunk space requirements depend upon the size of the cattle, and the type of pens. For feeder weight animals (500-1,000 lb) a minimum of 10" of bunk space per head should be provided. Twelve to 14" is better. Limiting bunk space causes cattle to eat less often but more at each feeding, which reduces performance, compared to eating the same amount over a longer period of time. The digestive system does not function as efficiently when feed is ingested in large batches vs. smaller more evenly spaced batches. In addition, more acidosis and founder will result from less frequent feeding. These recommendations assume ad libitum (all the cattle can eat) feeding. If limit feeding is practiced, as in a stockyard situation, bunk space must be increased to 24-36"; otherwise, the bigger more aggressive animals will eat a disproportionate share.

As for pen space, in open lots 200 sq. ft. per head is adequate under most conditions. If drainage is poor, more space will be required. In slatted floor confinement lots, 25 to 30 sq. ft. per head is often considered adequate. In the extreme Southwest (southern California, Arizona, New Mexico and south Texas), summer feeding performance will be enhanced if shades are provided. Shades should provide from 10 to 15 sq.ft. of shade per animal.

DEVELOPING RATIONS

RATION CONDITIONERS

To ensure maximum intake, processed grain rations must have an appealing texture. Texture is a very difficult term to define, but is reasonably easy to see when one looks at or feels a ration. In most cases, what ration conditioners do is very loosely bind the grain particles together so as to eliminate dust and noticeable fines.

The three most common ration conditioners are silage, molasses, and fat. Added singly or in combination they can enhance the texture of most processed grains. As a general rule, when silage is used as the source of roughage, other conditioners are not necessarily needed (but are often used). There are many ways to use conditioners, but the author's preference is to use silage with dry-rolled or dry heat-processed grain (popped), and molasses and/or fat with steam-flaked grains (and a dry roughage such as chopped hay). With high-moisture grain rations, molasses can often be used to improve texture.

GROWING RATIONS

Growing rations consist primarily of roughage in order to keep energy levels low. Grains are often added, but usually at levels of no more than 20-40% of the ration.

A large majority of the growing rations fed are based on corn silage. Although classed as a roughage, corn silage is actually a blend of roughage and grain, and the amount of grain can be quite variable. Silages made from varieties of corn designed for grain production can be as high as 45% grain, whereas forage varieties normally make silages consisting of 15 to 25% grain, and can be as low as 5 to 10%. Consequently, predicting gains for cattle fed corn silage based rations can be difficult.

In years when grain is cheaper on an energy basis than rough-age, up to 20% dry grain is often added to corn silage rations. This makes cattle gain faster and have more favorable conversions and costs of gain for the growing period, but seriously hinders performance during the finishing phase.

THE NEED FOR THE DUAL PROGRAM FEEDING METHOD

A fairly common practice among farmer-feeders is to hold cattle from growing weights to slaughter weights on a ration con-taining approximately a 50/50 mix of silage and grain. Either that, or hold them on a growing ration until they reach finishing weights, and then switch them to an approximately 50/50 - 60/40 grain/silage ration for finishing. The reason is usually that they market a silage crop through cattle, and this allows them to feed more silage than if they used a high grain finishing ration. Typically there are few records kept, and so the program appears reason-ably good.

Actually, a 50/50 grain/silage ration is inefficient. The pri-mary reason for inefficiency is that a 50/50 mix will not create the proper rumen environment for either amylolytic (starch using) or cellulolytic (fiber using) microorganisms. As a result, neither the silage nor the grain is digested efficiently.

Byers (1975) found that when a silage ration contained over about 40% grain (including grain in the silage), performance would fall short of what the net energy system would predict. Actual gain, of course, would increase due to additional energy (grain), but it would not be as great as the energy values themselves would predict, indicating a digestive inefficiency. For that reason, many nutritionists recommend using growing rations that contain no more than about 35% total grain (on a dry matter basis).

In addition to the inherent inefficiency of the 50/50 ration, holding cattle on that type of ration from grower weights further decreases total overall performance by making them too fleshy (fat) at too light a weight; i.e. if calves are put on that type ration at 400-500 lb, they will be carrying considerable fat by the time they reach about 700 lb. In order to be merchantable to most packers however, heifers must weigh 800-900 lb and steers 1,000-1,100 lb. Carrying cattle up another 300-400 lb when they are already fleshy increases gain costs substantially.

For the sake of efficiency, cattle should go through what is known as a two-stage or two-phase program; (1.) a genuine low energy growing type ration from weaning to feeder weights (for gains of no more than about 1.75 lb/hd/day). (2) a high energy finishing ration from feeder to slaughter weights.

DEVELOPING THE FINISHING RATION SERIES

As all professional cattlemen know, cattle cannot be put on high grain rations right after their arrival in the feedlot. Cattle that have been consuming primarily roughages must be gradually adjusted to high grain rations. Normally it takes about 3 weeks to bring cattle up to a high grain ration (about 90% concentrate).

Actually, it is not the animal that needs to become adjusted, so much as it is the microbial population in the rumen. When starch is introduced into the rumen, it is broken down rather quickly by the microorganisms. Lactic acid is given off as a by-product, and in animals that have been consuming mainly roughages, there will be a very low population of bacteria that can utilize the lactic acid. Thus, lactic acid can increase very rapidly in cattle abruptly moved to rations high in grain. The pH in the rumen of cattle eating roughages will be about 6.5; when abruptly moved to high grain rations, it can go as low as 3.5. When this happens, the animal suffers what is known as lactic acidosis. Acid enters the bloodstream and severely upsets the animal's metabolism. In acute cases the animal may stagger, fall, go into convulsions and die. In less severe cases, the symptoms of founder may appear. In founder, the animal's feet become extremely tender, making

Figure 6-21. Typical foundered hoof.

it very difficult for them to walk. This is because the hoof becomes engorged with blood, and through a mechanism not thoroughly understood, the 3rd phalanx bone of the foot rotates downward, creating severe pain and pressure upon the sole of the foot. Over a period of time the hoof will grow at an extraordinarily rapid rate from the toe, with distinctive annualer rings running the full length of the hoof.

Treatment consists of taking the animal off feed and drenching the rumen with mineral oil, which causes the digestive system to be purged, and to a certain extent, prevents absorption of nutrients (lactic acid). In acute cases the animal may be intraveneously injected with sodium bicarbonate. Of course, the best solution to the problem is prevention; newly arrived cattle brought up on grain rations gradually.

When designing a series of rations, or when bringing cattle up on feed, it should be remembered that the faster the cattle are moved to the "hot" ration (high concentrate ration), the better the performance will usually be. (Get the cattle off inefficient roughage/concentrate combination rations as quickly as possible.) Indeed, many feedlot managers believe that a certain amount of founder and acidosis is inevitable . . . "If you don't have any stiff (foundered) cattle, you aren't pushing them hard enough".

The idea then, is to move cattle onto grain rations gradually, but quickly. To do this, most feedlots utilize a series of approximately 4 rations; a top high concentrate ration, and 3 intermediate rations. In addition, most feedlots will feed the 1st or starter ration over hay for 2 or 3 days (increased roughage). A typical ration series using a dry roughage would look something like Table 6-4 .

Table 6-4. DRY ROUGHAGE FEEDLOT RATION SERIES.

	Ration			
Feedstuff	1	2	3	4
Grain	48	60	72	83
Dry roughage	45	33	21	10
Ration conditioner (molasses, fat, etc.)	3	3	3	3
Protein supplement	4	4	4	4
	100%	100%	100%	100%
Days fed	7-10	5-7	5-7	finish

Since the ration conditioner and protein supplement are considered concentrates, ration #1 becomes a 55:45 ratio of concentrate: roughage. Successive steps up to ration #4 are gradual increases in grain, and decreases in roughage. Ration #1 is fed longer than the other intermediate rations because sometimes it is several days before some cattle go on feed (begin eating substantial quantities of the ration). Cattle often triple their initial intake in the first few days, and it obviously would not be good if they were moved to a higher concentrate ration just as they began increasing their intake.

Flexibility is left in the number of days each ration is fed since some cattle can be moved quicker than others. Cattle that have never been in a feedlot before are usually frightened and often will not begin eating appreciable amounts of feed for a week or more. Cattle that have been held in stockyards with feedbunks, etc. often go right to the bunk and begin eating immediately. Obviously, they may be moved through the rations quicker than slower starters. Cattle with Brahman blood in them are more prone to acidosis than European breeds; old cows, particularly if in poor condition, must be moved slowly.

If high-moisture feeds are used in the ration series, then allowances must be made for the moisture. For example, silages normally run about 35% dry matter, whereas dry roughages such as hay normally run about 90% dry matter. If silage is substituted for hay, the amount of dry matter or roughage value of silage vs. hay must be determined. At 35% dry matter, silage has about 39% the roughage value of hay (.35 ÷ .9 = .39 or 39%). This means it will take approximately 2½ times as much silage as it does hay to get the same roughage value (dry matter).

In order to adjust a set of rations such as Table 6-4, a ratio

Table 6-5. HIGH MOISTURE ROUGHAGES SUBSTITUTION RATION SERIES.

	Ration			
Feedstuff	1	2	3	4
Grain	28	39	53	71
Silage	65	54	40	22
Ration conditioner*	3	3	3	3
Supplement	4	4	4	4
	100%	100%	100%	100%

*As a practical matter, whenever silage makes up the entire roughage portion of the ration, conditioners are usually not needed.

between grain and roughage must be established, and then the silage (high-moisture feed) adjusted for its moisture. Using the same grain:roughage ratios used in Table 6-4, the substitution of silage for a dry roughage would make them appear as shown in Table 6-5.

ADJUSTING CHEMICAL ANALYSES FOR MOISTURE

One of the most confusing aspects of ration formulation is adjusting and compensating for variations in moisture. All feeds contain some moisture. Dry feeds such as grain or hay usually contain from 8 to 15% moisture.

Laboratory analyses or evaluations of feeds are usually given on an as fed basis, a dry matter (100% dry) basis, or on an air dry basis. As fed means with the full amount of moisture; dry matter means with all the water removed; and air dry generally means with 10% moisture. Taking a sample of corn with 85% dry matter (15% moisture) and 10.5% crude protein on a dry matter basis, the analyses on an as fed, dry and air dry basis would vary as shown:

Corn

Reporting basis	Crude protein
As fed	8.9%
Dry matter	10.5%
Air dry	9.45%

A very striking difference is seen with high-moisture feed such as silages. A 35% dry matter corn silage (65% moisture) with 8% crude protein on a dry matter basis would thus appear:

Corn silage

Reporting basis	Crude protein
As fed	2.8%
Dry matter	8%
Air dry	7.2%

When comparing feeds then, it is necessary to know the moisture and the basis by which chemical analyses are reported. To adjust as fed analyses to dry matter, simply divide the figure by the fraction of dry matter; e.g. in the case of corn silage, 2.8% crude protein would be divided by .35 dry matter (2.8 ÷ .35 =

8.0%). To obtain the air dry basis, multiply the 100% dry figure by .9 (90% dry matter). Using the corn silage example, 8% crude protein would be multiplied by .9 (8% x .9 ≐ 7.2%).

LEAST COST RATIONS

The least cost ration concept has been around for a long time, but has been exploited much more since the availability of computers. The idea is that each available feedstuff is evaluated on a price per nutrient basis, and only the most economical feeds are included in the ration. The idea is basically sound, except that there are a number of associative inter-relationships between feeds that must also be considered.

While computers are often used to develop least cost rations, it is the author's opinion that simple arithmetic can often be just as quick. Only in areas such as California where numerous by-product feeds are available, is the computer a really valuable tool. For the most part, most feeders have no more than 10 or 12 feeds available, and their use can be worked out with a simple calculator rather quickly. For example, the typical cattle feeder may have the following feeds available at the following prices:

Corn - $89/ton
Milo - 78/ton
Barley- 85/ton

Alfalfa hay - $70/ton
Corn silage - 25/ton
Cottonseed hulls - 46/ton

Molasses - $90/ton
Feed grade fat - 360/ton

To determine the price per unit of energy value of these feeds, simply divide the dry matter content, and the energy value into the price. Using TDN* values from Appendix I the feeds would appear as shown in Table 6-6.

Of the grains available, milo appears to be the best buy at $111/ton of TDN. The feeder should remember, however, that the 78% TDN reported for milo is an intermediate value. Milo tends to be quite variable. In addition, as pointed out in Appendix I, the type and quality of processing the feeder has will make

*TDN chosen for simplicity. Net energy figures would be a little more accurate, but confusing to the first time reader. The concept, however, can later be applied with net energy values.

Table 6-6. PRICE PER TDN* VALUE OF SELECTED FEEDS.

Feedstuff	Price	÷ Dry matter (fraction)	= Price/unit dry matter	÷ TDN (fraction)	= Price/unit TDN
Corn	$89/ton	.855	$104/ton	.91	$114.38/ton
Milo	78/ton	.90	88.6/ton	.78	111.11/ton
Barley	85/tom	.89	95.5/ton	.83	115.06/ton
Alfalfa hay (28% fiber)	70/ton	.90	77.7/ton	.53	146.60/ton
Corn silage	25/ton	.35	71.42/ton	.65	109.89/ton
Cottonseed hulls	46/ton	.90	51.11/ton	.41	124.66/ton
Molasses (beet)	90/ton	.685	131.38/ton	.70	187.69/ton
Feed grade fat	360/ton	.995	361/ton	2.25	160.00/ton

*From NRC

a big difference. If steam-flaking, popping, or reconstitution is available, the 78% TDN value would be low, and thus milo would be the grain of choice. If dry grinding or rolling were the only processing methods available, the 78% TDN figure would be high, so corn would be the best choice.

For the roughages, corn silage is (and usually is) the cheapest on an energy basis. But as mentioned earlier, the associative effects of feedstuffs must be taken into consideration. If dry-rolled or popped grain is being used, the corn silage would make a very good ration ingredient (as the roughage). If steam-flaked or high-moisture grain is to be used, the alfalfa hay would be a better choice than corn silage. Cottonseed hulls are less expensive on an energy basis, but the protein provided by the alfalfa would still make alfalfa hay the better choice.

Of the ration conditioners, fat appears to be the more eco-nomical choice. Again, the other ration ingredients must be con-sidered. Fat usually should not be used as more than 3% of the ration, or reduction of intake will occur. Thus, if the grain is dry-rolled or popped and dry roughage such as alfalfa hay is used, some additional molasses would have to be used.

What the author has tried to do is explain how basic least cost values are obtained (arithmetically), while pointing out some of the extraneous factors that must be considered. In a good com-puter program these factors are put in as "constraints"; e.g. only 3% fat; no more than 6% molasses or 4% molasses and 3% fat, etc.

The computer may not, however, have all the "constraints" necessary to cope with an individual feeding situation. The feeder should therefore consider a computer formulated ration very care-fully before using it.

BALANCING RATIONS

THE PEARSON'S SQUARE

The Pearson's Square is the oldest and simplest method of balancing rations. When using two feeds, it will tell exactly the ratio of each feed required to meet the desired specifications. For example, assume it is desired to feed a milo ration using alfalfa hay as the source of additional protein. If the ration is to contain 11.5% crude protein (CP), and the milo has a CP value of 9.5%, and the hay 15%, the following arithmetic steps would be used:

Setting Up The Square

Set up the square by placing the desired CP level in the middle, and the CP values of the two feeds in two corners on one side:

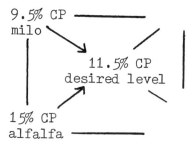

The next step would be to subtract the CP of the feeds across the desired CP level (one of the feeds must be higher than the desired level, and one must be lower).

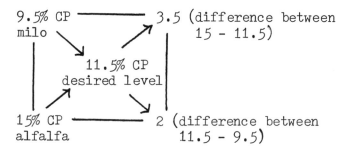

Now add the total of the differences to get the total parts of the two feeds to go into the ration:

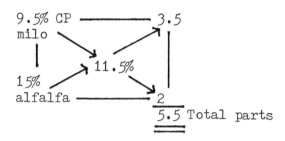

To get the percentage of alfalfa to go into the ration, divide the difference from the subtraction of the desired level from the milo, by the total. That is, divide the number diagonally across from the feed by the total. For alfalfa the percentage in the ration would be:

$$\frac{2}{5.5} = .36 \text{ or } 36\%$$

For the amount of milo, the percentage in the ration would be:

$$\frac{3.5}{5.5} = .64 \text{ or } 64\%$$

To check and see that 36% for alfalfa and 64% for milo are right, multiply times their CP value as a fraction, and add the results:

	% in Ration		CP (Fraction)		
Alfalfa hay	36%	x	.15	=	5.4%
Milo	64	x	.095	=	6.08
					11.48%

For a finishing ration, 36% alfalfa hay is too much roughage. Thus, the way most rations are actually balanced using the Pearson's Square is to first decide what the grain/roughage level is to be and then use a third feed as the protein concentrate. In this case, let us assume that a 90% milo, 10% alfalfa hay mix is desired, and cottonseed meal is to be used as the additional protein source. The first step then, would be to determine the protein content of the 90% milo, 10% hay mix:

Feed	%	x	CP Fraction	=	CP in Ration
Milo	90		.095		8.55
Alfalfa	10		.15		1.5
					10.05% CP

The Pearson's Square would then be used like this:

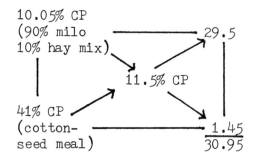

$$\% \text{ milo/alfalfa mix} = \frac{29.5}{30.95} = 95\%$$

$$\% \text{cottonseed meal} = \frac{1.45}{30.95} = 5\%$$

The only problem in using this method is that the final ration will not contain 90% milo and 10% hay. Rather it will have something less than those values (95% of those values). The ration would actually be:

	Original %	x	Adjusted fraction	=	Actual % in ration
Milo	90		.95		85.5%
Alfalfa	10		.95		9.5
Cottonseed meal	5		--		5.0
					100.0%

As a practical matter, the roughage level is the only change that need be of concern (to maintain a minimum level). In this case, the .5% could be added in at the expense of milo (change final ration to 85% milo and 10% alfalfa).

The author would also like to point out that nearly all feedlot rations can be made more economical by the addition of some urea or other NPN. Performance is also usually slightly better, since a moderate amount of urea will keep N readily available for the rumen microorganisms (Chapter 12). To add urea correctly to a ration (using Pearson's Square method of balancing), a set level should be added (rather than as a variable like natural proteins). To do this simply figure the urea into the ration mix (as was done with the milo and alfalfa), and then adjust with a natural protein concentrate; e.g. cottonseed meal.

FORMULATING FEEDLOT SUPPLEMENTS

The Pearson's Square is the method of ration balancing most commonly used in teaching. In actual practice, somewhat different methods are used since a separate premix supplement is typically used. The Pearson's Square can still be used, but it is not quite as accurate as other, somewhat more complicated methods. The procedure outlined here is only one of many procedures that may be used, but the author feels this one is probably the easiest to follow.

Protein is the item that is usually balanced first. In the past, crude protein has been considered an adequate protein evaluation for use with ruminants. The procedure has been to select the crude protein level deemed necessary (from NRC, etc.), and then balance for it. There are a number of ways of doing this, but the most common has been to; (1.) first determine the major feed ingredients (grain & roughage); (2.) determine the level of crude protein provided by those ingredients; and then (3.) determine how much additional protein must be supplied via a supplement. Usually, the percent supplement physically going into the ration is decided in advance by whatever level the feedlot feels it can mix adequately. For most feedlots, this level runs from 3 to 5%.

For example, assume that it is desired to feed a ration consisting of 80% dry-rolled milo, 10% cottonseed hulls, 5% molasses, and 5% supplement. The first step would be to determine the crude protein supplied by the milo, alfalfa, and molasses by multiplying their fraction of the ration times their crude protein value:

Feedstuff	Fraction of ration (% x .01)		Crude protein		Crude protein in ration
Dry rolled milo	.80	x	9.5%	=	7.6%
Cottonseed hulls	.10	x	4.0	=	.4
Molasses	.05	x	3.0	=	.15
	.95 or 95%				8.15%

If it is determined that the ration should contain 11.0% crude protein on the as fed basis, then the supplement must supply an additional 2.85% crude protein (11.0 - 8.15 = 2.85%) to the ration. The amount of supplement to be added to the ration has already been set at 5%. To determine the level of crude protein the supplement must contain, divide what it must supply (2.85%) by its level in the ration (5%). Thus, the supplement must contain 57% crude protein (2.85% ÷ 5% = .57).

One of the primary ingredients in nearly all cattle supplements is urea, since it is nearly always cheaper than natural proteins on a percent equivalency basis. Of course, the amount that may be used is limited. The rule of thumb used in the past has been a maximum of 30% of the total protein consumed. In actual practice, 20% is much more widely used in the feedlot industry today.

If 20% of the total protein as urea is deemed acceptable, this means that 2.2% of the ration may consist of urea equivalency (.20 x 11% = 2.2%). Since only 2.85% of the ration's protein needs to be supplied by the supplement, then 77% of the supplement's protein value may consist of urea equivalency (2.2% ÷ 2.85% = 77%).

It is important to understand the difference between percent protein equivalency and the percent urea in the supplement. As pointed out in Chapter 12, urea has a 281% protein equivalency. To determine the actual percent of urea that will go into the supplement, divide the amount of crude protein needed by the protein equivalency of urea (281%). Since the supplement must contain 57% crude protein and 77% of it can come from urea, then 43.89% protein equivalency can come from urea (57% x .77 = 43.89). Dividing 43.89% by 281% (urea's equivalency) reveals that 15.6% urea may be added to the supplement (43.89 ÷ 281 = 15.6%).

Now the natural protein source must be added. To arrive at how much natural protein must be added to the supplement, divide the amount of natural protein needed by the amount of protein equivalency contained in the natural protein selected. In this case 23% of the crude protein in the supplement must be supplied by a natural protein source (100-77% from urea = 23%). Multiplying .23 times the 57% crude protein (that the supplement must contain), means that 13.1% protein equivalency from natural protein is required (.23 x 57% = 13.1%). If the natural protein selected is cottonseed meal (with a crude protein of 41%), divide 13.1% (crude protein needed) by 41%. Thus, 32% of the supplement will be cottonseed meal (13.1% ÷ 41% = 32%).

Forty-seven and six tenths percent of the supplement is now known (32% cottonseed meal + 15.6% urea). This leaves 52.4% of the supplement for minerals and filler (grain by-products, etc.). To balance for minerals, follow the same procedure for protein. For example, to balance for calcium and phosphorous, first determine what is supplied by the ration:

Feedstuff	Fraction of ration	Ca	P	Ca added to ration	P added to ration
Dry rolled milo	.80	.03	.28	.024	.224
Cottonseed hulls	.10	.13	.06	.013	.006
Molasses	.05	.66	.08	.033	.004
				.070	.234

If it is desired to have .45% Ca and .28% P in the ration, then the supplement must add .38% Ca (.45 — .07) and .046% P (.28 — .234) to the ration. To determine how much Ca and P must be contained in the supplement, divide what is needed (.38% Ca and .046% P) by how much supplement is in the ration (5%). Thus, the supplement must contain 7.6% Ca and .9% P.

Before adding a Ca and P source, it should be determined how much is contained in the other ingredients in the supplement. Urea, of course, contains no minerals. Cottonseed meal, is rather high in P as it usually contains about 1%. Since cottonseed meal makes up 32% of the supplement, then it alone would bring the P content of the supplement up to .32% (.01 x 32% = .32%). Phosphorous must therefore be added to increase the supplement P content by .58% (.9% — .32% = .58%).

Defluorinated rock phospate is the most commonly utilized P source for feedlot supplements. Normally it contains about 18% P and 33% Ca. To add .58% P to this supplement, about 3.25% defluorinated phosphate will have to be added (.58% ÷ 18% P = 3.22%). This will also bring the supplement Ca rating up to about 1% (.0325 x 33% Ca = 1.07% Ca). This, of course, reduces the amount of Ca from another source that will have to be added. The supplement still needs to be 7.6% Ca, but since the addition of the defluorinated phosphate has added 1% to the supplement total, additional Ca will only have to be 6.6% overall.

Calcium carbonate (limestone) is the most common source of Ca. Normally limestone contains about 33% Ca and no P. Thus, 20% limestone will have to be added to the supplement (6.6% ÷ 33% = 20%).

The proportions of the major ingredients of the supplement are now known:

Ingredient	Contribution to supplement	Percent
Urea	Non-protein Nitrogen	15.6%
Cottonseed meal	Natural protein & some Phosphorous	32.0%
Defluorinated phosphate	Phosphorous & some Calcium	3.25%
Limestone	Calcium	20.0%
	Total	70.85%

This leaves 29.15% for trace minerals and filler. Trace minerals may be added similar to Ca and P; i.e. by examining what is contained in the ration. More often trace minerals are added in a standard package either to the specifications of a particular nutritionist, or as a commercial product marketed by a feed company.

PREDICTING FEEDLOT PERFORMANCE

Rough predictions of feedlot performance may be estimated by calculating net energy values or the older TDN values (Appendix I). For high concentrate rations, TDN values will predict performance nearly as accurate as net energy values, In high roughage rations, TDN values will usually overestimate performance.

The author would like to emphatically point out that subtle differences in any ration and/or feedlot management, can significantly reduce performance below what calculated estimates would predict it to be, regardless of the system used. These prediction systems assume ideal feeding conditions and do not take the associative effects of feedstuffs into consideration. The use of these systems should therefore be limited to teaching, other academic/ scientific pursuits, or as a check against actual performance to determine if optimum performance is being obtained. The development of cost of gain (performance) data for the purposes of hedging or other financial purposes, should come only from actual experience with the same type of ration, cattle and conditions as will prevail for the period to be projected.

Figure 6-22. Feedlot performance can be altered significantly by inclement weather (above), and/or inferior genetic quality of cattle (below).

Figure 6-23.

The reader should not interpret this to be a criticism of these systems because is is not. The scientists who developed the equations used in these systems were knowledgeable, and clearly the most recent system developed, the California Net Energy System, is about as accurate as can probably be developed. The only point the author would like to stress is that subtle differences can cause actual performance to vary significantly from projected data.

PREDICTING PERFORMANCE WITH THE TDN SYSTEM

The TDN system is the oldest and simplest of the performance prediction systems, and will be explained first because of its simplicity.

The first step is to determine the TDN content of the ration. To do this, multiply the percent TDN (from Appendix I) of each feedstuff in the ration times the amount (decimal fraction) of dry matter* in the feedstuff. For example, if a ration consists of 85% barley, 12% alfalfa hay, and 3% molasses, the TDN of the ration would be:

Feedstuff	TDN		Fraction in ration		Dry matter fraction		TDN added to ration
Barley	83%	x	.85	x	.89	=	62.7%
Alfalfa hay (24% fiber)	55%	x	.12	x	.88	=	5.8
Molasses (cane)	72%	x	.03	x	.77	=	1.7
							70.2%

The next step is to estimate consumption. In this example, assume that the cattle to be fed are 650 lb steers which will be marketed at weights of 1,100 lb. Using that ration, an average consumption of about 22 lb could be expected (from Table 6-2).

To determine how much total TDN the cattle will eat, multiply the total lb of ration times the decimal fraction of TDN in the ration (.702). In this case, the cattle would consume an average of 15.4 lb of TDN (.702 x 22 lb).

To determine the average gain, take the average weight of the cattle (875 lb) and look at Appendix Table II to see what cattle of the nearest weight would theoretically gain if they ate 15.4 lb of TDN. According to Appendix Table II, 882 lb cattle (ave. wt.) would gain about 2.6 lb.

---

*TDN figures given in the appendix are on a dry matter basis.

PREDICTING PERFORMANCE WITH THE CALIFORNIA NET ENERGY SYSTEM

The California Net Energy System is somewhat more complex than the TDN system since it differentiates between the ability of a feedstuff to simply maintain an animal, and the ability of the feedstuff to supply energy for growth or gain. Thus, the system places two separate energy values on each feedstuff; NEm, Net Energy for Maintenance; and NEg, Net Energy for Gain.

As with the TDN system, the first step in estimating performance is to determine how much energy is in the ration. In this case, two energy figures, NEm and NEg have to be calculated. The format for calculating the two figures is essentially the same as the format used to calculate the TDN in the previous section.

Feedstuff	NE_m (Mcals/cwt)		Fraction in ration		Dry matter fraction		NE_m added to ration
Barley	96.6	x	.85	x	.89	=	73.1
Alfalfa hay (24% fiber)	56.6	x	.12	x	.88	=	6.0
Molasses (cane)	83.3	x	.03	x	.77	=	1.9
							81.0

Feedstuff	NE_g (Mcals/cwt)		Fraction in ration		Dry matter fraction		NE_g added to ration
Barley	64.4	x	.85	x	.89	=	48.7
Alfalfa hay (24% fiber)	26.6	x	.12	x	.88	=	2.8
Molasses (cane)	53.3	x	.03	x	.77	=	1.2
							52.7

Thus, this ration contains .81 megcal/lb NEm, and .527 megcal/lb NEg. (Figures in tables and calculations express energy as megacalories per cwt.) To calculate gain it must first be determined how much of the ration the animal needs to satisfy its maintenance requirement. To do this, take the average weight of the animal and look at Appendix Table V to see what the maintenance requirement of that weight of animal is. If dealing with steers going from 650 lb to 1,100 lb, the maintenance requirement for 880 lb steers (average weight) would be about 6.95 megacalories of NEm. The ration contains .81 megcal/lb NEm.

To determine how many pounds of ration must be consumed to meet the maintenance requirement, divide the maintenance requirement by the NEm in the ration (6.95÷.81). Thus, the cattle must consume 8.6 lb of the ration to meet their maintenance needs. If consumption is 22 lb then there will be 13.4 lb of ration left over for gain.

Multiplying 13.4 lb of ration times the NEg contained in the ration determines how much NEg will be available for gain. Thus, there are 7.1 megacalories of NEg available for gain (13.4 lb ration x .527 megacalories/lb).

Looking at Appendix Table V, it is shown that 880 lb steers receiving 7.1 megacalories of NEg (left over after maintenance) will gain at a rate of between 2.8 and 2.9 lb. In comparison with the TDN calculations done in the previous section, there is a discrepancy of about .25 lb/day. The discrepancy can be explained by the fact that the California Net Energy System assumes that the cattle have been implanted with a growth stimulating hormone. Implants normally increase gain by about 10%, which would therefore increase gain about .25 lb.

In general, the California System is more accurate for prediction of feedlot performance. However, the author would again like to point out that given less than ideal weather, cattle of below average genetics, failure to allow for associative factors in rations, etc., the California System (as well as the TDN System) can significantly overestimate performance.

CHAPTER 7   BUYING AND SELLING CATTLE

Probably nowhere else is the philosopy of "caveat empor" (let the buyer beware) more applicable than in cattle trading. There have been attempts by the Dept. of Agriculture and other groups to standardize cattle grades, but as a practical matter there are so many variables that all cattle should be evaluated individually.

WEIGH UP

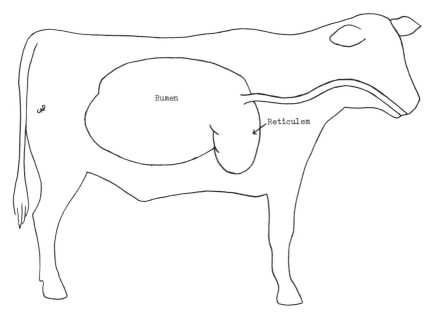

Figure 7-1. FILL POTENTIAL OF BEEF CATTLE.

The weighing conditions prescribed for a cattle trade have a tremendous impact upon the actual price that is paid. This is more true with cattle than any other type of livestock since cattle carry a larger percentage of their weight as ingesta, and do not have the water retention mechanisms that other species have. Under certain conditions cattle can carry as much as 30% of their weight as feed ingesta. This is primarily due to the great size of the rumen and the large amount of water carried in the rumen (Figure 7-1). Most ruminants have a water retention mechanism that retains fecal matter in the large intestine and absorbs most of the water before defecation. The fecal matter is rolled around in the large intestine, and when completed the feces take the form of dry hard balls of manure. Sheep, goats, deer, etc. have this mechanism, but cattle do not. Thus, with nearly 30% of the cattle's weight in ingesta and the inability to retain most of the water, cattle can lose a substantial amount of weight rather quickly when cut off feed and water.

This type of weight loss is known in the trade as shrink. Shrinks of up to 10% are quite common on cattle held off feed for as short a period as 24 hr. Under special circumstances, shrinks of up to 18% have occurred.

Cattle bought out of livestock auctions and stockyards will typically have a greater shrink than cattle bought directly from the producer. This is because auctions and stockyards manage their cattle so as to create a maximum "fill". The cattle are usually fed only very palatible hay which produces a maximum intake of both feed and water for the short period of time the cattle are there.

Experienced cattle traders often consider weighing conditions and shrink the entire purpose of cattle trades. For example, if a cattle trader buys a set of yearlings off a ranch after being gathered and driven for several hours, he can expect the cattle to shrink substantially, quite possibly 6%. Assuming a 600 lb pay weight at 85¢/lb, the buyer will pay $510/hd. If the cattle are held for 3 days and fed hay, and then sold for the original purchase price, a substantial amount of money will be made. Trucking costs on a 100 mile haul (to the trader's pens) might average $2.50/hd and feed costs a $1.00/hd/day. The cattle trader therefore has $515.50 per head in the cattle ($510 purchase cost + $2.50 trucking + $3.00 hay). However, when the cattle are sold, they will have gained back their 6% shrink and weigh 636 lb instead of 600 (600 x 1.06). Thus they will return $540.60 for a profit of $25.10/hd. On a trade of 500 hd, the cattle trader will have made $12,550 simply by selling the cattle for the same price he originally paid.

As mentioned earlier, cattle feces are very high in moisture. Thus, everytime cattle defecate, they lose from 4 to about 12 lb. Activity stimulates cattle to defecate and nearly every knowledgeable cattle trader takes advantage of this. Oftentimes cattle buyers insist on walking through pens of cattle for the sole purpose of "chousing" or "stirring" the cattle around (expressed purposes are usually to get a better look at the cattle, etc.). Another ploy often used is to demand that certain individuals be cut out of the herd or bunch. This, of course, causes the cattle to be stirred around, and can cause the seller to lose a substantial amount of money in a short period of time.

As a general rule, the thinner the cattle the greater the potential for fill and subsequent shrink (rumen capacity in relation to body weight). Likewise the older the cattle, the greater the fill and shrink potential (due to proportionately larger rumens). Old range cows with their angular body configuration and large paunches, have more potential for fill and shrink than any other class of cattle. Highly palatable roughages usually create the greatest fills since they stimulate the greatest intake of both feed and water. Grain rations typically do not create as much fill as roughages. However, on any given type of cattle and ration, fill and shrink can often be the spread between profit and loss.

PRIOR NUTRITION OF FEEDER AND STOCKER CATTLE

The prior nutrition of feeder and stocker cattle makes a great difference as to how the cattle will perform. As a rule, it is best to obtain cattle that have been deprived nutritionally, but are otherwise healthy. This is due to the phenomenon of compensatory gain.

The adage of compensatory gain says that the more an animal has been deprived, the more rapidly it will gain when provided for nutritionally. There is a point, of course, where the health of the animal suffers and increased death loss offsets the compensatory gain of the surviving animals. It is therefore not ordinarily desirable to purchase animals that are literally at the brink of starvation, but it is always desirable to have animals that are thin versus those that are fleshy. Thin animals can gain from 5 to 15% faster than comparable fleshy animals, and this response can greatly influence feed costs and profitability.

Cattle that have been fed grain should be discounted in relation to cattle that have not been fed grain. Be cautious of cattle represented to have been on growing rations, as substantial amounts of grain are often added to these rations. It is very common to see silage rations that contain 20% added dry grain. Twenty percent dry grain in a silage ration is actually about 45% grain in the ration on a dry basis (20% $\div$ 35% dry matter silage). In addition, silage usually contains 10-30% grain, making the so-called "growing" ration a relatively high energy ration.

During periods in which grains are cheaper on an energy basis than roughages, it is fairly common to see grower weight cattle fed high grain rations. Before being taken to livestock auctions, etc., grain fed calves or yearlings are sometimes fed a laxative roughage such as alfalfa for 3 or 4 days. Thus, when the cattle defecate the feces are green and watery, making the cattle appear to have been held on pasture.

PURCHASING CATTLE AT LIVESTOCK AUCTIONS

In the past, livestock auctions in some areas have had less than a shining reputation. While there have been some abuses, most auctions are run honestly and above board. Still, however, when purchasing cattle from an auction, the buyer should be cautious, and see who it is that bids against him, etc. When in a new area, it is often best to use the services of a reputable order buyer as he will be aware of how the local operation is run.

When buying from an auction the thing the buyer must be most concerned about is animal health. As explained in Chapter 4, most viral diseases have a 10-14 day incubation period and thus cattle (particularly calves) can look healthy, and then fall apart days later. The problem is greatest in auctions because traders often move cattle through several auctions in a matter of days before selling them. This not only stresses the cattle, but brings them into contact with a vast array of disease organisms. This is where a good local order buyer can be of assistance, as he will often know the origin of many of the cattle coming through the auctions in his area.

As mentioned earlier, cattle moved through auctions are typically fed for maximum fill. For that reason, in most cases auction cattle should not be purchased unless the price is a discount compared to what similar cattle could be purchased for in the country.

BUYING CATTLE THROUGH ORDER BUYERS

Selecting an order buyer is much like selecting a business partner. The man must be honest, intelligent, knowledgeable and have a sincere interest in supplying what you want.

When explaining what type of cattle you want, be very specific. Good buyers will honor all reasonable specifications. However, when specifications are unrealistic, they will often meet the price limitations, but quality or health of the cattle will suffer. It is therefore vitally important that you really and truly communicate with your buyer. In setting a price limitation, be sure that you ask if he feels the limitation is realistic. The situation order buyers dislike most is being told to buy the best quality, and best health, at a price substantially under the market.

When taking delivery of cattle it is always good business to weigh the cattle. The order buyer may have the purchase weights, and it may be an inconvenience to weigh the cattle, but always weigh the cattle on delivery. When order buyers have known that clients did not have scales, there have been cases of outright fraud. Certainly delivery weights will be less than purchase weights, but when shrink becomes greater than would ordinarily be expected for the length of haul, you will have an indication that it is time to find a new order buyer. Usually, when the order buyer knows the cattle are to be weighed on delivery, there will be no problem. Aside from keeping everyone honest, it is just good business to know the delivery weights or shrink on every load of cattle.

After the cattle are weighed, examine them carefully to see that the other specifications have been met. Look at each animal individually. A sometimes used trick (particularly with crossbred cattle), has been to meet the type and quality specifications on most of the cattle, but run two or three clowns* on the bunch. As an example, Mexican cattle have a reputation for being very healthy, hardy cattle. There are stories in the Southwest where Mexican cattle (particularly Brahman crosses) have had a scattering of southeastern cattle thrown on them, and the entire load represented to be Mexican cattle.

Cattle from the western part of the U.S. are generally considered to be genetically superior and healthier than southeastern cattle. Many times cattle operators specify (to order buyers) that their cattle must come from Montana, Wyoming, etc. There have apparently been cases where order buyers have purchased cattle in the Southeast and then shipped them to the West. When shipped a day or two later, the cattle inspection papers will now be by the state of Montana or Wyoming, etc. (Even without any intended fraud, there is a good chance the client will get southeastern cattle since cattle traders are constantly running southeastern cattle through western auctions.)

In short, when dealing with order buyers; look for integrity, be realistic, and always weigh cattle upon delivery.

SELLING RANCH OR PASTURE CATTLE

In selling cattle the owner must usually be very concerned with the weighing conditions. When dealing with cattle traders remember that they make their living by their wits. When making a deal, make sure all the details are specified and agreed upon. Again, be sure there is no doubt as to the weighing conditions.

If cattle are bought off of lush pasture and need not be hauled very far for weighing, quite often the buyer will specify a shrink,

*Lower quality cattle (in the vernacular).

say 5%. This, of course, is because the cattle will be full, and would lose considerable weight on a longer haul. Technically, this means that the pay weight is the actual weight of the cattle minus 5%. Thus, if the cattle weigh 600 lb and the terms are 75¢/lb with a 5% shrink, the pay weight will be 570 lb. The cattle will therefore cost $427.50/hd (570 lb x $.75). Shrink, of course, is simply a part of price. The buyer can offer a higher price but with more shrink, or a lower price with less shrink, and the net to the seller can be about the same.

Under ranch conditions where cattle must be driven long distances before being weighed, they will sometimes be sold without a shrink. In that case it must be remembered that if the cattle are held off feed and water, shrink can run into double digits. With weighing conditions like that, certainly a premium price would have to be paid.

SELLING FEEDLOT CATTLE

Selling fat cattle to packer buyers is usually a barter type transaction. There are a few packer buyers who are straightforward and will bid what the cattle are worth from the beginning, but usually it is a game of verbal gymnastics.

Packer buyers often will try and run down the quality of the cattle to the owner or feedlot manager negotiating the sale. One of the most common (and irritating) maneuvers is to say the cattle need to be on feed longer, say 30 more days. If the seller refuses to accept the bid, when the buyer returns in 30 days, the comment is often that the cattle are too fat.

The amount of shrink taken is usually standardized. In most of the feedlot areas of the U.S. a 4% shrink is standard. Thus, most of the bartering usually centers around price only.

Packer buyers will often insist that certain individual animals be left out of a trade, or that they be "sold on the rail". Selling on the rail means that a liveweight price is not specified, and that the animals are bought on a carcass basis; i.e. the packer pays for the cattle on a carcass grade and yield basis. Usually, this is done when the buyer is unsure of what the carcass of the animals will be like, or they otherwise do not fit the rest of the animals.

Selling large numbers of cattle on the grade and yield basis is a controversial practice. Many cattlemen's organizations feel that this is capitulating to the packers, giving them increased bargaining power over other feeders. However, at certain times, individual feedlots have no choice and cannot get packers to accept cattle on any other basis. Most commonly, this is when the individual feedlots are small and some distance from the packer.

At the time of writing, to eliminate this as well as other problems, many small operators in the Midwest have begun employing the services of brokers who represent their cattle to packers and negotiate the trade.

BUYING COMMERCIAL* BREEDING STOCK

Purchasing breeding stock is a situation which requires substantial planning as well as shopping. The first step, of course, is selecting a breed. Indeed, the importance of selecting a breed (or more often a particular cross) should not be underestimated. All too often tradition and fads have had undue influence on this decision (see page 23).

Under normal economic conditions, it is almost always better to acquire breeding females by raising replacement heifers from your own herd. One of the primary reasons for this is that it greatly reduces the chance of introducing diseases into the herd. Brucellosis (contagious abortion), as well as other venereal diseases can be introduced and spread into a herd by just one animal; the results can be disastrous if it occurs. If replacement females must be purchased from an outside source, it is a good practice to keep them separate from the rest of the herd and do not allow the bulls servicing them to come into contact with other females in the herd.

When purchasing females it is usually better to buy heifers than mature cows, as most cows offered for sale will have been culled in some way, most often for reproductive problems. Heifers offered for sale, of course, will be the ones that a cow-calf producer chooses not to hold back, but there is a better chance that they will be genetically and reproductively superior than mature cows. The exception to this would be dispersal sales where entire herds are sold off.

When purchasing heifers the buyer has the choice of going to breeders who specialize in raising breeding stock, or buying heifers sold through regular channels as stocker or feeder cattle. The question is one of paying a premium for breeding value, or paying the meat value for ordinary cattle. When paying extra for increased breeding value, the buyer must, of course, be certain that it is actually there.

The first decision that should be made when bulls are needed is whether to purchase bulls, or to purchase semen and initiate an artificial breeding program. Certainly the biggest advantage

*Non-registered animals for meat production (as opposed to registered animals raised for sale as breeding stock).

to artificial insemination (A.I.) is the availability of higher quality bulls than an individual cow-calf producer could afford to purchase. A reduced chance for disease introduction, is also an advantage to A.I. (when semen is purchased from established, reputable companies).

If the extra management necessary for an A.I. program does not seem justified and bulls must be purchased, they should be examined and proved free of venereal diseases before given access to the herd. Likewise, bulls should be fertility tested prior to purchase. Table 7-1 represents a study conducted by New Mexico State Univ. in which only 73% of the bulls in service were found to produce semen of satisfactory quality and quantity (Ruttle, 1980).

Table 7-1. SEMEN CLASSIFICATION OF NEW MEXICO BULLS BY AGE.

Age	total number	Satisfactory	Questionable	Cull
1	45	80	13	7
2	248	73	16	11
3	267	79	12	9
4	197	83	8	10
5	156	84	5	11
6	136	79	11	10
7	129	81	11	8
8	94	66	11	23
8+	48	35	19	46
Total	1320	73	12	15

When purchasing bulls, a tremendous variation in quality and price is usually available. Selecting a particular quality and price is a very difficult decision. The most difficult problem is assessing the quality. Once the buyer is certain of the quality, then it must be quantified; i.e. if the buyer knows a particular bull or set of bulls will increase the weaning weight of his calves by say 7.5 lb (page 238), then he can begin to determine how much he can afford to pay for the bulls.

BUYING REGISTERED BREEDING CATTLE

In the registered cattle business, promotion is the name of the game. During the so called exotic craze of the early 70's, Madison Avenue would have done well to have taken notice of some of the tactics used.

Putting all the advertising, popular press articles, and testimonials aside, evaluating registered cattle becomes a matter of interpreting data and the truthfullness of the people supplying the data. Certainly evaluating the animals themselves is important and must be done, but it can be deceiving. Moreover, the most valuable purpose in examining the animals themselves is to note the presence or absence of any gross physical faults (e.g. pendulous udders, double muscling, etc.). [Remember, cattle may not show a bad physical feature (such as dwarfism) but can carry the gene for it.]

Trying to determine by physical examination if a group of registered cattle will have increased weaning weights, etc. is at best, an educated guess. To begin with, there can be substantial genetic differences between similar looking animals. In addition, registered animals offered for sale are usually pampered and fed much better than commercial breeding stock are, the reason being to make them look better than they actually are.

The data most useful is usually weaning and yearling weights. Again, it can be expected that these weights will be favorably influenced by treatment and feeding superior to what most commercial cattle get. Most registered calves are either creep fed or run on irrigated or otherwise improved pastures, or both. Likewise, registered yearling cattle are typically "supplemented" on pasture. Most often the "supplement" contains several pounds of grain.

In the last few years a number of registered breeders have begun sending their bulls to centralized performance testing centers. Essentially, these centers are feedlots which feed standardized rations so that performance (rate of gain) between bulls can be evaluated. The data from these centers can be very useful, but again care must be exercised in interpreting it, as prior nutrition can greatly affect the performance. As explained earlier in the chapter, if animals are held on substandard nutrition levels for a period of time, they will gain significantly faster than animals held on adequate nutrition levels, when put on feed at a later date. For this reason, the data supplied by test stations is best used to evaluate bulls within a particular herd, rather than between herds.

All in all, when purchasing registered cattle one must be totally objective in examining both the animals and their data. At most registered auctions high pressure selling is usually the keynote. Whip popping ringmen and high flying auctioneers typically try to create an air of excitement. When attending these sales, the buyer should already have examined data on the animals, and know how much he can afford to pay for them. He should not be influenced by the prices animals sell for at the sale itself, as often there will be a few animals that seemingly sell for quite high prices . . . higher than their data might indicate they're worth.

SUMMARY

Quite obviously, cattle trading is not the kind of thing that can be learned out of a textbook. For that reason the material presented in this chapter has been rather general in nature, the main thrust being to merely point out some of the everyday tricks and pitfalls often practiced.

CHAPTER 8   SPECIAL TOPICS IN ECONOMICS AND ACCOUNTING—AS RELATED TO THE CATTLE BUSINESS

CATTLE CYCLES

The cattle business is one of the purest forms of free enterprise in the U.S. today.   For all practical purposes, it is a pure supply and demand situation.   As the supply of cattle goes up, cattle prices typically drop, and vise versa (Figure 8-1).

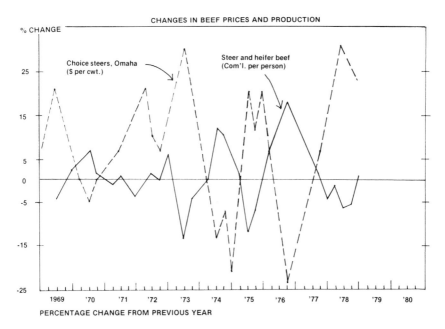

Table 8-1. CHANGES IN BEEF PRICES AND PRODUCTION.
Source: USDA

Since about 1925, cattle numbers in the U.S. have gone through cyclical fluctuations about every 10 years (Figure 8-2).   That is, cattle numbers have reached an extreme high and low about every ten years.   These cycles have caused equally extreme fluctuations in cattle prices.   Because of this, the cattle business in the United States has typically been a feast or famine business.

Essentially what happens is that when cattle prices go up, producers hold more heifers for breeding and cull fewer cows. As a result, cattle numbers increase until supply exceeds demand and the price goes down.   With declining prices producers then begin holding fewer heifers and cull their cows more heavily.   Cattle numbers thereby decrease until a "shortage" occurs and prices begin to increase. The situation is usually accentuated by financial institutions which typically increase their lending when the cattle market is good, and decrease lending when it is bad.

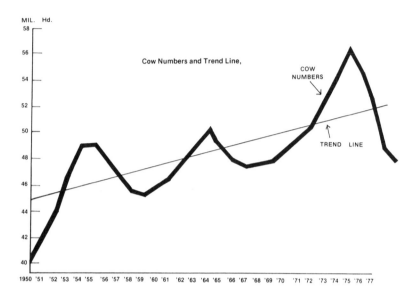

Table 8-2. CATTLE SUPPLY CYCLES.
Source: Cattle Fax, National Cattlemen's Asso., Denver, CO

The cattle cycle has its greatest effect upon cow-calf producers since cow herds are long term investments. Feedlot and stocker operators are, of course, greatly affected by the cycle, but they are also greatly affected by short term market fluctuations since they buy and sell every few months.

The National Cattleman's Association has considered attempting to make all cattlemen aware of the cattle cycle in the hopes of controlling it. The idea being that when cattle numbers begin to increase excessively, it could be brought to the attention of the nation's cattlemen. Ideally they could begin selling (culling) the cow herd off early and hopefully avoid a major market break. In the author's opinion this is going to be an extremely difficult task.

HISTORICAL CATTLE CYCLE WARNING SIGNALS

by Tommy Beall
Economic Research Director, Cattle Fax
National Cattleman's Association

The cattle cycle is probably the most important long term aspect of the cattle business. The true cause of cycles is the industry's bottom line. If most cattlemen are showing profits, there will be herd expansion. If most are losing money, liquidation is inevitable.

We have a unique business. In order to decrease supplies in response to market signals, the industry first has to further increase beef production through herd liquidation. To increase supplies, the industry temporarily has to decrease output further, in order to build up the basic "factory" (hold back heifers for rebreeding).

Most businesses have cycles of some type, but most cycles are shorter than ours. Weather plays an important role in the cattle cycle. Because beef supplies are highly dependent on forage and grain production, we are closely tied to weather cycles.

There has tended to be a serious drought every 20 years, and these droughts have been factors in our most severe liquidation periods - in the mid-1930's, 1950's and 1970's. If history repeats itself, we will have another serious and widespread drought and another severe liquidation period, in the 1990's.

Weather cycles affect beef supplies and costs, while other related cycles - such as the hog cycle and business cycles - affect the demand for beef. All affect cattle prices and/or profits.

Now, what are some of the signals we should watch for in getting ready for the cattle cycle's next liquidation phase and prolonged period of lower prices?

REGULARITY OF CYCLES

Because the last several cycles have occurred with regularity, one signal to consider is simply time. Since the 1930's, each liquidation period has begun in a year ending in "5" - 1945, 1955, 1965, and 1975.

The liquidation phase of each cycle has tended to last 2 to 4 years. Expansion periods have lasted 6 to 8 years. Once we reach the end of the liquidation phase and the industry becomes profitable, it takes 1 to 3 years to materially increase beef supplies and 5-6 years to reach a point of over-expansion. This assumes that demand follows previous patterns and grows along with population and income.

As we monitor the cycle, there are two basic supply factors to watch: (1.) Total cattle numbers and the cow herd. (2.) Total slaughter and beef production as compared to potential slaughter and beef output at a given time.

CATTLE AND COW NUMBERS

Historically, annual growth of 1 to 2% in total cattle numbers has not caused an over-supply problem. Anytime we have had sustained growth in cattle numbers of more than 2% a year, the industry has over-produced. Output has grown faster than dollar demand, and cattle have been sold only at loss levels.

The same signal applies to the cow herd. A sustained cow herd growth rate of more than 2% always has got the industry into trouble.

An early indication of excessive growth in the cow herd is found in data on replacement heifers. A normal replacement heifer holding rate is about 20% of the cow herd. When heifers kept for herd replacement exceed 21% of the cow herd, the industry usually has run into over-production.

SLAUGHTER DATA

Data on cattle slaughter provide one of the best indicators of where we are in the cycle. If the herd size is to remain stable or grow very slowly, the industry has to kill 34-35% of the total inventory each year. In 1973, we slaughtered only 30% of our Jan. 1 inventory. This was one reason for over-production. Then, to get numbers back into better balance with demand, we had to slaughter about 38% of our inventory for three years afterward.

This slaughter rate of 35% of the inventory is close to what we call the potential kill each year - numbers equal to the previous year's calf crop plus death losses. The difference between actual kill and potential kill represents either herd expansion or herd liquidation.

FEMALE SLAUGHTER

Cow slaughter at a rate less than 15% of the total cow herd indicates herd expansion is occurring. Cow slaughter at less than 13% of the cow herd indicates the industry is headed for over-expansion.

Changes in heifer slaughter reflect the extent to which heifers are being held for replacement or are being slaughtered or fed out for market. When the cattle herd is growing at a "normal" rate, 26-27% of our slaughter is heifers. In the early 1970's heifer slaughter averaged 25-26% of the kill. From 1976-1979, during liquidation, it averaged almost 29% of total slaughter.

Total female slaughter (including cows as well as heifers) of less than 85% of steer slaughter, has also indicated that the industry was increasing its breeding herd too fast.

OTHER FACTORS

In addition to monitoring supply factors, we must look at demand and at production cost factors.

Other indicators to watch:

1. Cost of gain compared to fed cattle prices. When cost of gain went above fed cattle prices in 1974, calf prices went from $10 over the fed market to $10 below the fed market in one year.

2. Feeder cattle prices in relation to the beef market. We are getting into liquidation when feeder cattle prices go below fed prices and packers are able to buy and kill young steers and heifers as non-fed beef.

3. Number of cows per 100 persons. This figure relates potential beef supplies to demand. The equilibrium point - meaning a steady and reasonably good market - in recent times has been 24 cows per 100 persons.

4. Incomes and spending on beef. If the public spends an average of 2.5 to 2.6% of its disposable income for beef, the industry can remain in fairly good shape for some time. This assumes incomes continue to grow at least as rapidly as industry costs increase.

5. Per capita beef supplies. In recent years the industry has tended to remain profitable when per capita supplies have not exceeded about 115 lb per year on a carcass weight basis. When the total was above 120 lb, supplies were excessive.

No one signal tells the whole story of the cycle. But, if we keep an eye on all of the indicators mentioned here, we will have a better idea of what the future will bring. Keep in mind, however, that these have been historical relationships.

In previous cycles we have had a stable economy and a stable demand. Supply has been the main variable. Going into the 1980's we cannot assume that the economy and demand will remain stable. Some of the historical relationships we have looked at in the past are changing--mainly due to inflation and sharply higher production costs. The historical "signals" still must be monitored; but due to cost and demand changes, other factors are becoming more important than supply relationships.

FORWARD CONTRACTING

Forward contracting is like most any other business contract in which a seller agrees to deliver a specified product or commodity at some point in the future for a specified price. Seldom are breeding cows sold by forward contract. There are a limited number of fed cattle contracted by packers, but the purchase of calves or yearlings by stocker operators or feedlots is relatively common. In the case of yearlings, the cattle are often inspected by the buyer and then contracted for a later delivery. With calves, the contract will usually call for all or part of the calf crop of an entire ranch or farm.

The terms of the contract will vary, but most contracts will obviously specify the cattle, date of delivery, and price. Of utmost importance are the weighing conditions (page 138), as this can greatly influence the actual selling price. In addition, management practices of the cattle before delivery should be spelled out. Of particular importance is any type of grain feeding (pasture supplementation, creep feeding, etc.), as this will increase the weight of the cattle and decrease their future gain performance.

As with any type of business transaction, it is important to know the reputation, integrity, and financial backing of the other party to the contract. Aside from the usual avenues of fraud possible with livestock (switching cattle, cheating on weighing conditions, altering scale tickets, etc.), during adverse marketing conditions, there have been a number of failures to execute contract. When cattle prices go up substantially, there have been cases of contract sellers reporting abnormal death losses, etc., or just outright refusing to deliver (selling to someone else at the higher market price). Likewise, when cattle prices go down substantially, contract buyers have been known to not honor contracts. Of the problems associated with forward contracting, failure to deliver during a rising market is probably the most common. Therefore it is wise to have liquidated damages (specified penalty) for failure to execute contract written into the agreement.

HEDGING WITH THE FUTURES MARKET

To many cattlemen, particularly in the Midwest, the futures market is a very controversial subject. The main objection to the futures market has been that it has allowed big business to get into the cattle business, and indeed, some very large firms have entered cattle feeding and use the futures market extensively. Many complaints about the futures market have resulted from attempts to use the market without fully understanding how it works, which has resulted in serious financial problems (explained later in text). In addition, there are some who believe that the futures market allows speculators to control the cattle market.

While controversial in the eyes of many cattlemen, the futures market is here to stay. Like it or not, the futures market does have a bearing upon the cattle industry, and so it is necessary for cattlemen to learn a few basics whether they plan to use the futures market or not.

WHAT FUTURES PRICES ARE

Futures prices are quoted for live cattle and feeder cattle contracts. The price is given on a hundred weight basis; e.g. $70/cwt, but the total price is for an entire contract.

Live cattle contracts consist of 40,000 lb (live weight) of choice quality slaughter steers (approximately 37 steers weighing 1050-1100 lb). Feeder cattle contracts call for 42,000 lb of choice quality feeder steers weighing approximately 650 lb each (about 65 steers).

WHAT FUTURES PRICES ARE NOT

Futures prices are not indicators of what the market will be. On the contrary, in some cases the futures market has had an inverse relationship to what the actual market really was, particularly with feedlot cattle. To be specific, there have been instances where the futures price for live cattle advanced to what seemed to be inordinately high levels. Seeing this, cattle feeders responded by placing large numbers of cattle on feed timed at finishing during the months with the correspondingly high futures prices. As a result there was a surplus of slaughter cattle available at the time of delivery, and the actual market price was quite low compared to what the futures price was several months earlier (Purcell 1977).

THE MECHANICS OF HEDGING

In hedging, the producer sells one or more contracts* to coincide with the number of cattle he will have for sale (or wishes to hedge). The delivery month is selected to coincide with when his cattle will be ready for market. In essence, he sells a contract in which he promises to deliver the type of cattle indicated during a specified delivery month at a set price. The contract is bought by another

---

*The contract amount (40,000 lb, "live cattle" and 42,000 lb, feeder cattle) are designed to be what one semi-trailer truck can haul. Just before going to press, half contracts (20,000 lb) were offered for live cattle.

individual, usually a speculator.* The commodity exchange does not buy or sell any contracts itself, but acts as an agent, providing a place for transactions and rules to govern those transactions.

When the delivery month arrives, or the contract seller otherwise feels it necessary to liquidate his contract(s), the usual procedure is to buy an offsetting contract(s). That is, if he has contracted to deliver live cattle, he simply buys a contract(s) for the same month. Having sold and bought the same number of contracts, they nullify each other. If he wishes, instead of buying back his contract(s) he may deliver cattle to correspond with his contract(s). Less than 3% of the cattle sold on the futures market are actually delivered (CME 1978), but the option is there.

Live cattle	Par	Feeder Cattle 50¢ Discount	75¢ Discount
Peoria & Joliet, Ill.	Omaha, Neb.	St. Paul, Minn.	Billings, Mont.
Omaha, Neb.	Sioux City, Iowa	Greely, Colo.	
Sioux City, Iowa	Okla. City, Okla.	Kansas City. Kan.	$6.00 Discount
Guymon, Okla.		St. Joseph, Mo.	Montgomery, Ala.
		Dodge City, Kan.	
		Witchita, Kan.	
		Amarillo, Tex.	

For a true hedge, the cattleman will sell his cattle and buy back his contract(s) simultaneously. The difference between the actual price of cattle in his area and the futures price during the delivery month is known as the basis. Usually the futures price and the local price (paid by packers, etc.) will be reasonably close. Most commonly, the difference will be whatever the trucking and other expense would be in delivering the cattle. However, if a disparity arises and the current futures price is significantly higher than the local price, the producer can deliver. It is the potential for delivery that keeps the futures and local price in line during the delivery month.

FIGURING BASIS

"Basis" is the difference between the futures price during the delivery month and the actual cash price. Normally futures prices will be slightly higher than actual local cash prices due to delivery costs.

---

* The speculator is often maligned by producers as being a "leech", etc. In reality, the futures market could not function without speculators as they provide liquidity. Without the extra capital (buying power) the speculators provide, it would be difficult to get in and out of the market in a short period of time.

In order to place an intelligent hedge, the producer must have a good idea what the basis or price spread will be. Usually this is arrived at by looking at historical data.

Assume that after careful study a producer decides that $2.00/cwt is a conservative estimate of basis for his area. If he therefore determines that $65/cwt would yield him a reasonable profit on his cattle, he must then sell his futures contracts for $67/cwt.

LIQUIDATING A HEDGE

In a true hedge the producer sells his cattle and "lifts" his hedge (buys back an equal number of contracts) at the same time. Assume that the producer has sold his futures contract for $67. This means that if necessary he may deliver live cattle meeting the contract specification and receive exactly that price.

More often, he will buy an equal number of contracts for the same delivery month he has sold his contracts for. The futures market will have fluctuated from whatever the price was when he placed his hedge several months back. But regardless of whether the futures price is higher or lower, he will net about the same amount of money . . . if he has figured his basis accurately.

For example, if he sold his contract in April for August delivery at $67/cwt, and when August comes the futures price is $62/cwt, then he will make $5/cwt when he buys back his contracts. The cash market however, will be close to the current (delivery month) price. In this example, assume $60/cwt ($2.00/cwt basis). Thus, since his target price was $65, then he will in fact receive that price. The transactions would be:

```
Transaction #1 (April) sell August contracts @           $67.00/cwt
Transaction #2 (August) buy August contracts @           -62.00/cwt
     Result of Transactions #1 & 2                        + 5.00/cwt
Transaction #3 (August) sell cattle on cash market @       60.00/cwt
     Net sale price of cattle                             $65.00/cwt
```

Assume now that the August futures went up instead of down. Instead of being $62/cwt, suppose it went to $75/cwt. If basis price has been calculated accurately ($2.00/cwt below futures in month of delivery), the producer will still receive a net of $65/cwt for his cattle:

```
Transaction #1 (April) sell August contracts @           $67.00/cwt
Transaction #2 (August) buy August contracts @           -75.00/cwt
     Result of Transactions #1 & 2                        (-8.00/cwt)
Transaction #3 (August) sell cattle on cash market @       73.00/cwt
     Net sale price of cattle                             $65.00/cwt
```

Notice that obtaining the target price ($65/cwt) was dependent upon calculating the correct basis. For this reason it is wise to be conservative in estimating basis.

BASIS RISK

Basis risk is a common term used in hedging. A literal definition would be the risk of the actual basis being greater than the anticipated basis. In actual practice, however, basis risk means different things to different people.

A simple example of basis risk can be shown by using the previous example in which $65/cwt was deemed to be a target price that would yield a reasonable profit. Anticipating a $2.00/cwt basis, the cattle were hedged at $67/cwt. There is a risk that the actual price spread between the futures price (during the delivery month) and the local cash price could be $3 or $4/cwt. If basis worked out to be $3 or 4/cwt, then the producer would only net $64 or $63/cwt for his cattle, which could force him into a break-even or loss situation.

As mentioned earlier, if the basis becomes too high, the producer has the option of delivering the cattle and collecting the face value of the futures contract he originally sold. However, in many cases the cattle raised and/or fed (particularly in the Southwest) are not deliverable, and thus basis risk becomes a much greater risk factor. Futures contracts specify choice grade cattle, and to deliver cattle on a futures contract, they must meet the specifications. Many of the cattle raised and fed (and hedged) do not meet the specification. In particular, the Brahma cross cattle which are so common in the Southwest, are usually not deliverable. Likewise, the large exotics (Charolois, Simmental, etc.) that grow quite well, but do not marble particularly well, are often not deliverable for live cattle contracts. When hedging these types of cattle, the producer must give great concern to estimating basis and be cognizant of the fact that he will have a substantially greater basis risk (than with British breed cattle, etc.). (Some cattlemen even hedge heifers, by figuring the basis difference between local heifer prices versus futures steer prices.)

MARGIN CALLS

Whenever a futures contract is sold or purchased, the entire value of the contract is not paid for, but the seller or purchaser is required to put up a margin. The actual amount will vary with the brokerage house, but 5% of the value of the contract is the usual minimum. As the futures price fluctuates from the original contract price, the contract holder will either get part of his margin back, or will have to put up additional margin money, depending upon which way the price fluctuates. For example, if a producer sells (hedges) a live cattle contract at $67/cwt, the contract would be worth $26,800 (40,000 lb x $.67). At a 5% margin rate, he would have to put up $1340. If the futures price went down to $65/cwt, the producer would receive part of his margin money back. This is because to liquidate his contract, he could buy a contract cheaper (9% cheaper) than the one he originally sold. If the price goes up, say to $69/cwt, then he must put up extra margin money to cover the price spread between what it would cost to buy a contract to cover (liquidate) the cheaper contract he sold. In this case, an additional $800 would be needed since a $69 contract is worth $27,600 and the $67 contract is only worth $26,800.

When extra margin money is required of a contract holder, he receives what is known as a "margin call". The purpose of margins and margin calls are to maintain the financial integrity of the futures market, but there is something inherently demoralizing about receiving margin calls. When margin calls come, they must be attended to quickly, and they can require substantial amounts of money. It is not uncommon to see 3 to $4/cwt price swings in one week. Just one pen of feedlot cattle (120 hd) would require 3 contracts (40,000 lb each) and with a $4/cwt price swing from $67 to $71/cwt, the contract holder would have to come up with $4,800 . . . quickly. A cattleman feeding out an entire calf crop of say 1,200 hd, would have to come up with $48,000 . . . just as quickly. There is nothing in the futures market that has caused as many problems for cattle producers as margin calls.

THE CLASSIC "WRECK"

In the U.S. there are individual cattle producers who are vehemently opposed to use of the futures market and are replete with stories of financial disasters caused by hedging. Usually these "wrecks" have been caused by unexpected margin calls (failure to prepare for margin calls).

The typical situation has been that cattle feeders unfamiliar with the futures market would sell several contracts as a hedge. The futures price would then rise rather quickly, creating margin calls for a substantial amount of money. Not having set up an account or line of credit for margin calls, the contract holder was forced to go to his banker for additional funds. The usual procedure with most financial institutions is for loan applications to be reviewed by a board of directors or loan committee which meets periodically. But margin calls must be met quickly . . .

at some point then, the contract holder was either not able to obtain the necessary funds without going through the usual channels, or was cut off completely.

Not being able to meet his margin calls, he was forced to liquidate his contracts. This meant he had to buy contracts at the current higher price to offset the contracts he originally sold as a hedge. In so doing, he was forced into taking a loss in the futures market while lifting his hedge and leaving his cattle unprotected. As mentioned earlier, whenever the futures market goes up substantially, oftentimes a larger than normal amount of cattle go on feed and drive the market down during the delivery month. In many situations then, the ill advised "hedger" also took a loss in the cash market.

PLACING A "GOOD" HEDGE

The most commonly cited rules for good hedging are:
1. Know the cost of production.
2. Figure basis objectively and conservatively.
3. Know how much risk can be assumed.
4. Set up a line of credit for margin calls before hedging.
5. Be a hedger - not a speculator.

A producer obviously cannot place an intelligent hedge unless he knows what his cost of production is. The idea behind placing a hedge is to lock in a profit. If the cost of production is not known, then the producer really doesn't know what price/cwt constitutes a profit, and he can conceivably lock in a loss. This may sound childishly basic, but there are still large numbers of cattle producers that do not keep detailed records and do not accurately know what their costs are.

Once the cost of production can be estimated with reasonable precision, the producer can determine what price/cwt (target price) will yield him an acceptable profit. To the target price, the basis must be added. Estimating the basis must be given at least the same consideration as estimating cost of production, because it is essentially part of the producer's breakeven figure. That is, if breakeven sale price is determined to be $64.00/cwt and basis is estimated at $2.00/cwt, then the producer would anticipate $66/cwt to be a breakeven hedge (futures) price. He knows (or thinks he knows), that he must wait until the futures market moves to $67/cwt or more, before he can hedge. Obviously then, if basis is actually $4/cwt, the producer would lock in a loss at $67/cwt.

Anyone using the futures market should analyze just how

much risk they can afford to take for two basic reasons. The first reason is that while hedging may lock in a profit, it also limits profit. If the cash market rises considerably, the hedged investment will make no more money than what it was sold "on the board" for (it may actually make less money due to interest on increased margin money). Many large cattle operators hedge only a portion of their cattle. What proportion should be hedged, and what proportion should be left open will primarily depend upon the net worth of the owner. The greater the net worth and income, obviously the greater the risk that can be assumed.

The second reason is that oftentimes when cattle are bought and placed on feed, they cannot be hedged profitably (for the month of delivery). Usually, at some point during the feeding period, the futures price for the delivery month in question will increase to where the cattle can be hedged profitably. A common practice among cattlemen that can afford the risk, is to go ahead and buy the cattle, and then wait for a favorable rise in the futures market. A study conducted at Oklahoma State University showed that even during the infamously poor market conditions of 1973 and 1974 there were very few instances in which at least a break-even hedge could not have been placed (Ikerd, 1977).

Whenever going into a hedging program always establish an account or line of credit for margin call purposes before any contracts are sold. Most financial institutions will establish a line of credit without charging any interest until the money is actually used. As mentioned in the previous section, failure to prepare for margin calls can place the hedger in an extremely unfavorable position.

As a final note, most authorities stress that a hedger should not be a speculator. When using the futures market as a hedging tool, stay with it. There is always the temptation to lift the hedge early (before the cattle are sold) and take a quick profit (leaving the cattle unprotected). Then too, after following the futures market for a period of time, one often gets a feeling of competence in predicting the market and there is a temptation to openly speculate (buy and sell contracts without the intention of hedging a given set of cattle). Speculation is, of course, the exact opposite of hedging . . . the object of hedging is to reduce risk, whereas speculation increases risk. [According to the Chicago Mercantile Exchange, 90% of all speculators lose money (CME,1978).]

CHAPTER 9   CATTLE AND INCOME TAX

Before beginning to evaluate this chapter, it must be realized that documentation is extremely difficult.

Personal tax returns are not a matter of public record, and high income taxpayers are unwilling to divulge their genuine motives. Indeed, if any information is given, it is usually misleading. The reasoning is obvious; the Internal Revenue Service (IRS) will disallow deductions from an "investment" operated for purposes other than profit (Internal Revenue Section #183).

Academic research into this area has been lacking. Taxation appears to have a relatively low priority for agricultural economic research, and certainly the difficulty of obtaining information from individual investors has also contributed to the lack of creditable literature.

Needless to say, tax shelters do exist . . . attorneys devote their lives to tax work, investment brokers sell limited partnerships with prospectuses that state "a change in the tax laws could significantly reduce the attractiveness of this program", and urban high income groups own cattle that they have never seen.

The purpose of this chapter is to provide general information only. The tax advantages of cattle ownership have been a significant factor in the U.S. cattle industry, and thus the author feels it necessary to provide the reader with background information. The contents of this chapter should not be used for decision making in tax strategy as sufficient detail is not presented (and is beyond the scope of this text).

THE BASIC TAX LAWS AFFECTING THE CATTLE INDUSTRY

To understand the sheltering effects a beef cattle investment may have, one should have an understanding of the basic laws around which most shelters are designed. A discussion of these laws, in effect negates a common misconception that tax shelters center around "loopholes"; i.e. it is commonly believed that tax shelters center around flaws in the tax code which allow unscrupulous lawyers and accountants to develop plans to keep wealthy clients from having to pay their "fair share" of income tax. In actual fact, tax shelters center around laws designed to promote investment . . . the effect of which is known to every lawyer, accountant, and knowledgeable congressman.

Likewise the common term "tax dodge" is a misnomer as tax dodge implies tax evasion which is a felony and punishable by imprisonment. Tax shelters, on the other hand, are investments which are given special tax treatment by postponing or reducing the tax liability which would otherwise be encountered without the investment; tax dodge refers to some kind of misrepresentation, which is fraud.

Of the laws affecting agricultural tax shelters, probably the most significant and certainly the most controversial, is that of cash accounting. In cash accounting the taxpayer (farmer) is allowed to report his expenses and receipts only as they occur; i.e. there are no adjustments for inventory. This is contrasted to accrual accounting in which inventory must be included.

For example, if a cattle feeder has been feeding cattle but has not sold the cattle prior to January 1st, under cash accounting all the feed and management costs up to Dec. 31st may be deducted. Since the cattle had not been sold, they need not be included in the accounting. Thus, the cattle feeder may report a loss. Under accrual accounting the cattle must be reported as inventory and an allowance made for them.

The effect of this provision is obvious; a taxpayer may postpone the recognition of income by not selling his crop until the following year. So in effect, farmers may even out their income between good and bad years.

Probably the second most important series of laws affecting agricultural shelters in general are the capital gains laws. Clearly, these are laws intended to increase total production as special tax treatment is given to "items purchased for the production of income". There are a number of details affecting the final tax rate, but for all practical purposes capital gains rates are 40% the ordinary rates.* Livestock are specifically included, although the holding period for long term capital gains for livestock is two years versus one year for other items.*

Accelerated depreciation makes up a concept covered by a complex of laws which greatly influences investment for qualifying items. For instance, section 179 allows an additional 20% depreciation, which in effect is a deduction. In other words, on a $20,000 piece of equipment the taxpayer may deduct $4,000** in extra depreciation the first year. If the taxpayer is in a 50% bracket then he has saved $2,000. Other laws allow increased depreciation rates over the standard straight line method (cost divided by useful life).

Increased depreciation rates reduce depreciation taken in later years, but under most circumstances financiers will advise

*As of 1980.

**$4,000 is the maximum that can be taken on a joint return.

early deduction because of the time value of money. With, say a 14% reinvestment rate, $2,000 saved a year from now would only be worth $1754 today (by the concept of present value). Saved 3 years from now, $2,000 would only be worth $1,350.* Thus, it is usually to the taxpayer's advantage to take deductions early.

A fourth law which affects a number of different industries is the "Investment Credit" law. Under this provision a taxpayer receives a credit against his income tax liability for 10% of the purchase price of a qualifying asset; i.e. a dollar for dollar reduction. In other words, if an investor buys $10,000 in farm machinery, he will pay $1,000 less in income tax that year.

Not to bias the reader, but at this point it should become clear that there is more to the Tax Code than the collection of revenue . . . the Federal Tax System is an integral part of fiscal policy.

There are a number of other laws affecting the material which will be presented later on in this chapter, but of these laws, cash accounting, capital gains, accelerated depreciation, and investment credit form a nucleus about which most agricultural tax shelters are centered, and therefore form a basis for an understanding of the mechanics of tax shelters. Other laws will be explained later as they are specialized and thus the context surrounding the necessity of their mention, will greatly facilitate explanation.

REASONS BEHIND EXISTENCE OF THE TAX LAWS ALLOWING AGRICULTURAL SHELTERS

According to Woods (1975) there are two schools of thought as to why farm related tax shelters exist: (1.) the tax laws were originally designed to simplify accounting methods for unsophisticated farmers, but are now being abused by wealthy "gentlemen farmers"; (2.) the tax laws were deliberately intended to subsidize agriculture.

The author feels that there is a third school of thought in which the main belief is that the existence of the tax laws are due to intense lobbying. It is this author's opinion that in reality the existence of the tax laws allowing sheltering of income are a combination of lobbying and direct subsidization, although the subsidization is probably intended for a broad scope of industry rather than just agriculture itself.

---

*The concept of present value takes into account the amount of interest or other monetary return a sum of money could earn in a given time period. It expresses what a sum of money would be worth today versus some time in the future (for a given interest rate). That is, money in the future is worth less, since it does not have the opportunity to earn the interest that money today has.

It is generally the contention of those upholding the theory of subsidization that the creation of the 50 to 70% marginal tax rates was done clearly as an incentive for reinvestment; i.e. to force wealthy taxpayers to reinvest their money and avoid such high nominal rates. While evidence of this is purely coincidental, it would appear to be true as it has been reported that effective tax rates rarely exceed 40%, and average 20 to 40% (Coats, 1974). As a 70% rate would seem confiscatory (at least to the author), it would appear that there is some connection with high rates and the written laws which allow effective reduction of those rates, while simultaneously stimulating investment.

As for the theory expressing belief that the laws are archaic, only one major law might fall in this category; Cash accounting. It is true that the cash accounting rule was originated in 1915 for accounting expediency and remains unchanged (Harrison, 1972), even though most farmers now utilize accountants for income tax purposes. But at the same time it cannot be said that Congress is unaware of the ramifications of that statute as cash accounting has been one of the most debated tax questions (involving agriculture) in the Legislature. All the other major laws have been modified somewhat since their origination. Clearly if these laws have been discussed and/or modified, then it cannot be said that Congress is not aware of the use of these laws for other than accounting expediency. As for lobbying, the author is not aware of any reference available in the literature, although many knowledgeable people believe lobbying has definitely been a factor.

At this point it is hoped that the reader will understand that the discussion concerning the rationale behind the tax preference statutes has not been an attempt to bias the reader one way or another. A value judgment as to whether tax shelters are good or bad has not, nor will be implied or inferred. The intent of this discussion has merely been to point out that Congress is aware of favorable tax treatment afforded by various tax laws. Indeed, coverage of the impact of these laws has been absolutely inadequate for quantification of any kind of a decision concerning the usefulness of tax shelters to society as a whole.

HOW CATTLE FEEDING TAX SHELTERS WORK

The principle behind cattle feeding tax shelters is to postpone income through the use of cash accounting. In essence, it is a method of averaging income. In a model study conducted by Willett et al (1974), the income averaging method provided by the IRS was considerably unfavorable compared to the deferral of income allowed by feedlot tax shelter programs.

In order to be able to take advantage of tax shelter programs, it is generally concluded that the taxpayer needs to be in at least a 50% tax bracket. Thus, when he expenses funds, it really only costs him 50¢ on the dollar.

Since feed may be deducted, but not cattle, the conditions that most favor cattle feeding shelters are when cattle are cheap and feed high. At the time of writing, cattle were high and feed relatively cheap. Still, however, there is room for shelter.

Assume an investor in the 50% tax bracket purchases 100 hd of 650 lb steers at 90¢/lb on October 1st. Cattle purchases are usually made with borrowed funds, so on a 30% equity basis the investment would look like this:

Investor's Equity	Bank Loan	Cost of Cattle (100 hd x 650 lb x $.90/lb)
$17,550	+ $40,950	= $58,500

Figure 9-1. SEASONAL INDEX OF FEEDER CATTLE PLACEMENTS, ALL GRADES AND WEIGHTS.

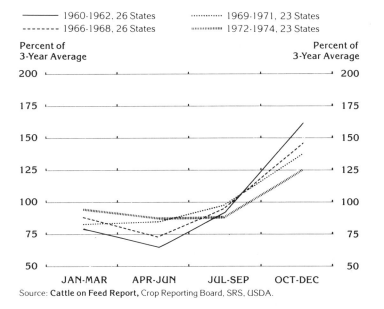

Source: **Cattle on Feed Report,** Crop Reporting Board, SRS, USDA.

Assuming a six month note at 12% interest, the interest charge would be $2,457.

If the cattle are placed on feed October 1st (note how feeder cattle placements increase in the fall - Figure 9-1), all the feed and management expenses are deductible up to January 1st (or whenever the investor's fiscal year for tax purposes ends). Assuming a 150 day feeding period and a 400 lb gain at $45/lb, total cost of gain would be $18,000 ($180/hd). The feed expense from October 1st to January 1st (120 days) can be expensed, and so 80% (120 days/150 days)* can be deducted. Thus $14,400 is deducted. Since the cattle are not sold until after January 1st, they are not reported on the tax return showing the feed expense. In this way, because cash accounting is allowed for "farming" enterprises, the investor reports a loss on his tax form, even though he has not sold the cattle yet. That is, he reports and deducts 80% (120 days/150 days) of the feeding and interest expense (if actually paid out).

Farming Loss

$14,400   feed & mgt. expense (18,000 x 80%)
  1,965.60 interest expense (2,457 x 80%)
$16,365.60 total loss for the year

Being in a 50% tax bracket, this saves the investor $8,182.80 (16,365.60 x .5). This is a true saving since he would have paid out that much in taxes.

The thing that bothers the "legitimate" cattle feeder is that the tax shelter effect lowers the investor's breakeven sale price; i.e. he can sell his fat cattle below the price necessary for an ordinary breakeven.

Without the tax effect, the breakeven sale price of the cattle would be the original cost of the cattle, plus feed expense, plus interest:

	Total Costs	Per Hd Costs
Original purchase cost of cattle	$58,500.00	$585.00
Interest expense	2,457.00	24.57
Feed costs	18,000.00	180.00
Total costs	$78,957.00	$789.57

*Feed expense would not occur evenly throughout the feeding period. But for ease of explanation it is assumed so in this example.

As 1050 lb steers, the cattle must return 75.20¢ ($789.57 ÷ 1050 lb) per pound in order to break even. But because the tax savings is actually a form of income to the investor, he has a lower break-even. His breakeven would consist of the total cost minus the tax saving:

	Total Costs	Per Hd Costs
Total costs (from previous example)	$78,957.00	$789.57
Less tax saving (from earlier computation)	– 8,182.80	– 81.83
Total costs for investor	$70,774.20	$707.74

As 1050 lb steers, the cattle need only bring 67.4¢ per pound ($707.74 ÷ 1050 lb) for the investor to break even. In other words, what would be an $83.96 per head loss to a "legitimate" cattle feeder, would be a breakeven situation for the 50% tax bracket investor. Likewise, what would be a meager $20/hd profit for the professional cattleman, would be over $100/hd profit for the tax investor. Professional cattlemen therefore feel that the non-agricultural investor has an unfair advantage.

When the cattle are sold in the following year, they are reported as income for that year. The investor, of course, can go through the same procedure and "roll over" the income with another group of cattle in the same manner, or utilize some other form of tax shelter. Even if he does nothing, he will have had the use of his tax dollars for a full year and the interest or return from whatever investment he used the money for. In addition, with inflation running in double digits, he can repay his tax dollars with cheaper dollars; i.e. at 14% inflation, by postponing taxes the investor will pay them with 86¢ dollars a year later.

Prior to the Tax Reform Act of 1975 the investor could capitalize on cash accounting even further. At that time the investor could borrow the money to cover feed expense and prepay both feed, management, and interest expenses. This really made cattle feeding attractive by allowing the investor to charge everything off, and do it all on borrowed funds so that he didn't reduce his own personal liquidity. In many cases, investors were able to save more dollars in taxes than they actually put up as equity.

At the present time, feed, management, and interest expenses may not be prepaid, and only personal equity may be deducted. In addition the IRS ruled that "farming investment" must not "materially distort income". At the time of writing the "material distortion" principle was just beginning to be tested in court.

IMPACT UPON THE CATTLE FEEDING INDUSTRY

Nowhere in agriculture have the tax laws had such a profound impact as they have upon the cattle feeding industry. In most agriculture the farmer/rancher sees the tax shelter investor as some type of villain. In this case, however, the feedlot industry went out and actively solicited the non-agricultural investor. As explained on page 85, in the late 60's the feedlot industry outstripped the availability of capital from the local rural banks. Big city banks were not interested in loaning on investments they knew nothing about, several hundred miles away. Of necessity, the feedlots turned to big city investors for their capital, and found a very salable commodity in the form of tax shelter feeding arrangements.

This literally revolutionized the feedlot industry. In the beginning, the feedlot industry came into being because it could feed cattle more economically than the small farmer feeder (by being able to afford costly feed processing equipment). With the advent of the tax shelter investment program, the success of a feedlot became much more dependent upon the ability to market securities rather than the ability to feed cattle economically. As feedlot capacity approaches 5,000 hd, an economy of scale is achieved (yard becomes more efficient due to its size). After about 15,000 hd a diseconomy of scale generally appears. In order to be able to afford the expense of registering and marketing securities (cattle feeding investment programs), the feedlots had to expand to at least 30,000 hd capacity (Meisner, 1974).

"Growth in feedlot capacities from December 31, 1970 to June 30, 1973 was almost exactly equal to the number of cattle owned by outside investors" (Simpson, 1974). In 1972 a special Tax Committee formed by the feedlots reported that the number of investor cattle doubled from 1970 to 1971, and then doubled again in 1972. At that time investor cattle accounted for half the cattle in the large yards and 20% of all fed cattle (Rhodes, 1973).

The infamous cattle market crash of 1974 brought an end to the rapid growth of the tax shelter feeding industry. Just how much the market crash was magnified by tax oriented feeding is difficult to say. Clearly, however, the rapid increase in demand for feeder cattle was a modifying factor.

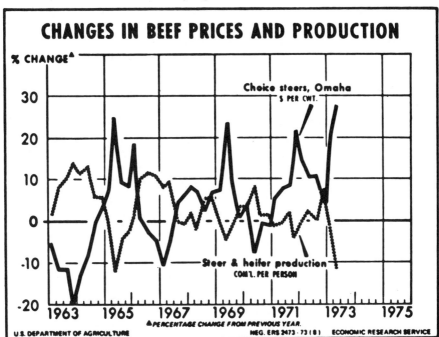

Figure 9-2. This graph was published by USDA just prior to the crash
of '74. Historically cattle prices had been a pure supply and demand
relationship (p.148). The thinking at the time was that we had entered
a new era . . . .

Cattlemen and investors alike took huge losses. The losses
suffered by a majority of the investors were in excess of what
could be offset by tax savings.

Some of the investors had purchased investment plans that
limited their losses, and were able to recover a good part of their
capital. At least one of the feedlots that had these provisions,
took bankruptcy.

A number of tax code provisions have been altered which
detract from the tax advantages of cattle feeding. These provisions
and remembrances of the 1974 crash have reduced the amount
of tax oriented feeding, but still it remains a definite factor within
the industry.

HOW CATTLE BREEDING TAX SHELTERS WORK

Cattle breeding tax shelters are extremely complex, and to
date many of the finer points have apparently not been tested

in court. The basic game plan is one of holding back heifer calves and raising them for use in the cow herd; i.e. the expenses involved in raising the heifers are charged off against ordinary income, but when they are later sold as cows (after a two year holding period), they are taxed at capital gains rates (about 40% the ordinary rate). In addition, the taxpayer is entitled to investment credit and accelerated depreciation on the purchase of the original breeding herd.*

The usual modus operendi is for only the steer calves to be sold each year; all the heifer calves are withheld. This generates a large **loss** which can be used to offset non-agricultural income while building up the taxpayer's net worth. Prior to 1969 the IRS could disallow the yearly loss deductions if they could prove that the business was not entered into for profit (IRS Section #183). This, of course, created a considerable amount of litigation. In 1969 that provision was amended so as to eliminate confusion (litigation). The provision now (as of 1980) states that if a profit is shown in 2 of 5 years the business will be assumed to be conducted for the purposes of profit. What is not made clear is if the profit must be "substantial"; i.e. can 3 years of large losses be offset by 2 years of minimal profits? At the time of writing this question had apparently not been tested in the courts.

Because of the two year holding period required for capital gain treatment, the investor is obviously "locked in" to the program for several years (so that heifer calves from several calf crop years can be converted to capital gain). In addition, because the IRS will disallow capital gains treatment for breeding stock raised for sale, the investor is limited to probably no more than two or three capital gain conversions (sell-off's of breeding cattle). That is, if an investor continually sells breeding animals the IRS will contend that he is in the business of raising and selling breeding stock, rather than calves. Additional sell-offs might be possible by waiting for drouths which regularly occur in the Western U.S., and contending that the dispersal was forced due to the drouth,

---

*Prior to 1976, taxpayers who used cash accounting and reported over $25,000 in farm losses, and had more than $50,000 in non-farm income, were required to keep what was known as Excess Deduction Accounts. The purpose of these accounts were to show the amount of farm losses (over $25,000) used to offset non-farm income. In years where farm profits were shown, the profits could be used to reduce the amount of "excess deductions" contained in the account. Upon sale of farm assets (such as breeding stock), which would ordinarily be capital gain items, they were taxed at ordinary rates to the extent of whatever was in the Excess Deduction Account. After Dec. 31, 1976 the law was eliminated, but taxpayers who had such accounts are required to maintain them until they are eliminated through profits or recapture of capital gain.

etc. The only problem there being that during drouths, cattle prices are typically depressed.

Because of the "lock in" and lack of liquidity for cattle breeding shelters, it is primarily investors with incomes that are continually high each year (such as physicians) that are attracted to these investments. Investors with incomes that fluctuate are more attracted to cattle feeding shelters because of the income averaging effect.

A rather unique aspect of cattle breeding shelters is that many investors use them as life insurance policies. They simply consider the yearly losses (from selling only the steer calves) as premium payments, while developing a large net worth (cow herd), for their heirs. The cow herd is built up by expensing ordinary income, but upon death, the heirs pay only inheritance taxes (about ⅓ the ordinary income tax rates) on the herd. Life insurance payments would, of course, also be taxable only at inheritance rates, but the premiums are not deductible. However, with the cow herd, the premiums (expenses) are deductible as business expenses . . . thus the investor has a life insurance plan with deductible premiums.

In actual practice the advantage of such a program over a normal life insurance policy is a question of time. It takes a number of years to build a cow herd up (increase the cash value of the policy) and so the question is quantified by deciding how far the investor is from death, and then plugging in the cost values; i.e. term life insurance values remain static (except for inflation) and so it follows that if the investor dies relatively soon, he may be better off with conventional life insurance.

TAX ASPECTS ASSOCIATED WITH RANCH REAL ESTATE

While there appear to be substantial tax advantages to the ownership of cattle, there do not appear to be significant tax advantages to the ownership of agricultural land. This is not to say that there are not economic advantages to the ownership of land, but rather that the intent may be inflation hedging or speculation. As early as 1968 (Martin and Gatz) it was reported that ranch properties (in Arizona) were bought for the purposes of "conspicuous consumption" or "ranch fundamentalism".

Regardless of what the actual motive is, due to the rapidly appreciating market values of ranch properties, it is clear that agricultural return (return from livestock) is not the motive. Due to the recent fluctuations in cattle prices it is difficult to arrive at an "average" return per cow (calf), but certainly $75 would be a maximum, if not optimistic average. Using the Capitalization Ratio Theory at an interest rate of 12%; the investment equation would be:

$$\frac{\$75 \text{ return}}{.12 \text{ interest}}$$

The maximum capitalization would therefore be $625. Yet it is common knowledge that ranch properties cannot be bought for less than $1,000 per animal unit, and often run as high as $3,000. Clearly then, there is some motive for purchase other than ordinary income from the ranch enterprise itself. Looking at sales data gathered throughout the state of Oregon, Eisgruber (1976) found that there was a general disparity between the market price and the agricultural value for all agricultural lands, but found the disparity to be two to four times as great for ranch lands than arable lands.

Rangelands bringing a higher market price per agricultural income value (over farm lands) would seem logical as there are far more tax advantages to livestock than there are to farm crops (Breimyer, 1975). But when one evaluates the total advantages, it would seem that there are no real advantages to the ownership of land . . . only livestock.

Prior to 1962 this was not the case as agricultural land did have a potential for conversion of ordinary income into capital gain, although there was an element of ambiguity within the law which resulted in a fair amount of litigation. The problem centered around land development costs (brush control, reseeding, etc.) and the Internal Revenue Service holding the view that farming has 3 distinct stages; (1.) preparation; (2.) development or pre-productive stage, and (3.) harvest. Preparation expenses must be capitalized; i.e. spread over a period of years. But once the preproductive stage is over (harvest begins), the expenses may be deducted currently (Davenport, 1975). The litigation with respect to cattle ranching was "Does grazing constitute harvest?". Operating on the premise that grazing does constitute harvest, high income investors could buy up large unimproved (brushy) ranches, invest large sums of money into brush control, fertilization, and seeding . . . deduct all this in the year of expenditure, and then later sell the ranch and be liable only for capital gain taxes on the increased value of the land.

In 1962, Congress severely curtailed opportunities for capital gain conversion with agricultural lands by passing the Land Clearing Section (Int. Rev. Sect. #182). Under this provision land clearing expenses may be deducted only to the extent of $5,000 or 25% of taxable income derived from farming only.

To further reduce the tax incentives to land clearing (range restoration) and ranch land in general, Congress passed in 1969 a resolution calling for recapture as ordinary income on land expenditures on farm land held for less than 10 years (Int. Rev. Sect. #1251).

The author is of the opinion that these provisions along with inflated market values have essentially eliminated income tax advantages to ranch lands. The basis of the author's opinion is the fact that the firms that sell or manage tax shelter breeding cattle investments typically lease or subcontract the land the cattle are run on. In some cases the firms may own the land themselves, but the author is unaware of any large scale tax shelter investment plans that include ranch or pasture land as part of the purchase.

This is not to say that investors are not purchasing ranch lands, because they are. However, the purchases are by individuals (rather than investment programs). It is the author's opinion that these purchases cannot be for ordinary income purposes (from cattle raising), or for tax shelter purposes. It is the author's opinion that these purchases are usually incidental to cattle ownership with a combination of motives consisting of speculation, recreation, and pride of ownership.

IMPACT UPON THE CATTLE BREEDING INDUSTRY

It has been reported that only 5% of the nation's breeding cattle are owned by non-agricultural forces (Woods, 1975), whereas at one time as much as 20% of the cattle on feed are owned by non-agricultural investors (page 168). Clearly then, tax induced investment has not had as great an effect upon the cattle breeding industry as it has the cattle feeding industry.

Cow-calf operators, however, have long complained that tax induced investments have caused ranch and pasture lands to be bid up beyond what a "legitimate" cattleman can afford to pay. As pointed out in the preceeding section, tax shelter motives are apparently not responsible for the inflated prices of ranch lands. Speculation on appreciation, inflation hedging, recreation, and pride of ownership are evidently the motives behind the purchases. Still, the purchases continue, and "legitimate" stockmen are resentful of urban investors paying more for property than they can afford to pay.

But while "legitimate" cattlemen are resentful of this situation, it should be realized that outside investors are not entirely to blame for the situation, as they represent only one side of the market . . . they are the demand. The existing landholders constitute the supply . . . a supply that is extremely limited due to an unwillingness to sell. Thus the situation becomes one of great demand and little supply and therefore the price is quite high.

So while the investor's zeal for land may appear to be uneconomic, so also is the existing landholder's unwillingness to sell. Most ranches have an average return to capital of about 4% (Oppenheimer, 1974), with most ranchers living very modestly (most of the revenue goes back into the ranch business).

Most properties would bring about $2,000/animal unit. Thus, for a 500 cow ranch, the existing landholder could sell out and receive a purchase price of about $1,000,000 - probably about $800,000 after taxes. Placed as a certificate of deposit at 9% interest, the rancher could receive $72,000 per year and never work another day of his life.

The crux of this whole discussion is simply to point out that there are more significant factors than tax considerations that are appreciating the value of range and pasturelands.

It would seem (to the author) that with the magnitude of tax induced cattle feeding that there is, the feeder cattle market would have to be affected. Obviously, this affects the cow-calf producer. While the cattle feeder may complain that investor demand for feeder cattle increases feeder prices, the cow-calf producer can hardly complain about higher calf prices.

As for the breeding cattle market, one must again consider that non-agricultural sources own only 5% of the Nation's cow herd. Also, since the nature of breeding cattle tax programs is to buy small herds and then increase the size of the herd for capital gain conversion, it is unlikely that outside investors significantly affect the commercial breeding cattle market; i.e. of the 5% owned by non-agricultural investors, probably only 2 or 3% were purchased.

There is a possibility, however, that the registered (purebred) cattle market may be significantly affected since most shelter programs deal in registered cattle, and the registered cattle business makes up a small percentage of the industry as a whole. The affect would probably be increased prices because of a greater demand due to the ability to offset high prices with depreciation deductions. This would be an advantage to the purebred breeder already in the business, but could act as an obstacle to a newcomer wanting to get into the business.

As mentioned in the beginning of this chapter, documentation of this type of subject is extremely difficult. Particularly difficult is attempting to assess the effects upon the industry. One might conclude that increased demand for cattle would be a benefit for the cattle breeder, but at the same time, may also make the cattle market more volatile. However, there is at least one benefit that a few cattlemen have enjoyed (due to tax shelter investors) that the author can point out by observation.

The apparent lack of tax advantages of owning land, coupled with the substantial tax advantages to the ownership of cattle, has created a demand for ranchers and farmers willing to care for (pasture) investor owned cattle. That is, without tax advantages to ownership of land, the investor is apparently better off financially to lease, rent, or otherwise contract land for his cattle since he can deduct lease payments or management fees, but not principle payments (on land).

It has been said that tax law has been developed too piecemeal for the creation of any central policy (Carman, 1975). But this situation does seem to create a workable symbiosis between legitimate cattlemen and wealthy investors, since one of the big drawbacks to the cattle business is the constant fluctuation in the cattle market. Fluctuation makes it very difficult to develop reliable cash flow budgets, and thereby reduces the certainty of the ability of the cattleman to repay loans. This, of course, reduces the amount of capital available to him. But instead of running cattle he owns, some cattlemen (ranchers in particular) have been able to manage investor owned cattle. Having a contract calling for a set fee substantially reduces risk, and thus this situation can and has made a number of ranchers more attractive borrowers. In discrete instances that the author is aware of, individual ranchers have been able to expand their operations in this manner.

CONCLUSIONS

Outside investor capital has had a significant effect upon the cattle feeding industry. It increased the size of a substantial number of feedlots beyond what would ordinarily be an economy of scale. It can exert an upward pressure on feeder cattle prices, and may possibly make the overall cattle market more volatile.

The effect upon the breeding cattle business is not as clear. The total tax oriented investment in breeding cattle is not as great as feeder cattle, and so the effect is probably not as great. There may be a significant effect upon the purebred business since most breeding cattle plans marketed as securities utilize pure-bred cattle. This is probably due to the increased value of pure-bred animals and the increased depreciation that would result. The inflation of rangeland prices is not related to identifiable tax motives. Land ownership is apparently incidental to cattle ownership, and inflated prices appear to be due to inflation hedging and "ranch fundamentalism".

The author does not mean to imply that the use of cattle invest-ments for tax shelter purposes should be either condoned or con-demned. Rather, the inclusion of this discussion has been to simply make the reader aware of a significant factor affecting the cattle industry. This is a subject an owner of a cattle operation will come into contact with, and so the author has attempted to provide a background of information. The information presented, however, has been general in nature and in no way specific enough for tax planning purposes.

PART II — REFERENCE

CHAPTER 10   BASIC ACCOUNTING AND FINANCE

The following chapter is intended for the use of readers who do not have formal training in accounting and finance. Those that are versed in these subjects will find the chapter painfully basic, but should realize that a substantial percentage of those involved in cattle production are not conversant in the areas of accounting and finance.

The subject matter will be limited to the explanation of three of the most common accounting and financial analyses; (1.) The Balance Sheet or Financial Statement; (2.) The Income Statement; and (3.) The Cash Flow.

THE BALANCE SHEET OR FINANCIAL STATEMENT

The Balance Sheet or Financial Statement is an accounting analysis that nearly everyone has had some kind of association with. Its purpose is to show the "balance" of assets versus liabilities and thus the name Balance Sheet. By noting the value of assets versus liabilities, the financial strength of the business can be seen, and thus the synonym, Financial Statement.

Table 10-1.   EXAMPLE BALANCE SHEET.

Goinbroake Feedlot
Tuffluck, Texas
Dec. 31, 1979

Assets		Liabilities	
Current Assets		Current Liabilities	
Cash on hand	$7.11	Bank loan on cattle	$397,853.02
Cattle	499,800.00	Accounts payable	121,473.21
Grain inventory	29,670.00	Wages payable	7,312.87
Silage inventory	194,250.00	Income taxes payable	576.18
Protein supp. inventory	3,315.00	Interest payable	78,418.12
Accounts receivable	24,410.09	Total current liabilities	$565,633.40
Total current assets	$779,452.20		
Fixed Assets		Long Term Liabilities	
Land	$24,000.00	Note on land (due 1987)	17,500.00
Pens & facilities	41,050.00	Note on constructed facilities (due 1991)	134,900.00
Feedmill	92,850.00	Total long term liabilities	$152,400.00
Feed trucks	23,248.17		
Office bldg. & equip.	24,872.16	Owner's equity	$297,439.10
Water wells & rights	30,000.00		
Total fixed assets	$236,020.33	Total Liabilities	$1,015,472.50
Total Assets	$1,015,472.50		

Assets refer to factors that are positive for the business. Anything that has a dollar value to the business would be included as an asset. Examples would be cash, machinery or other items that can be sold, accounts receivable, etc.

Liabilities refer to debts that the business may have. Examples would be unpaid bills (accounts payable), loans that must be repaid, etc. Also, the owner's equity is a debt that the business owes to the owners, and thus it is included with the liabilities.

Owner's equity is essentially the liabilities subtracted from the assets. Owner's equity therefore is an indicator of the financial strength of the business. A typical balance sheet is shown in Table 10-1.

In this balance sheet or financial statement it can be seen that the owner has $297,449.10 in equity or about 29% of the value of the company. When borrowing money, this is one of the first things a banker will look at as it gives him some idea of the financial strength of a business . . . the ability of the owner to repay loans. The greater the owner's equity, the greater the financial strength of the business. Probably the second thing most bankers look at is what is known as "liquidity", which is defined as assets that can be turned into cash rather quickly.

When evaluating liquidity, the first thing that is looked at, of course, is cash. In this case, the cash on hand is extremely low. While exaggeratedly low in this example, agricultural concerns do typically have little cash on hand. Usually what funds they have are tied up in the business.

The second thing the banker will look at is assets that can be sold quickly; i.e. liquid assets. In this example, nearly $500,000 in cattle are available. Cattle, of course, can be sold very quickly. The problem is that at certain times it may be unprofitable to the owner to sell them; i.e. when the cattle market is down, or the cattle are half way through a feeding period, etc. It should be remembered, however, that bankers are not really concerned with whether their borrowers make a profit or not. All they are actually interested in is the ability of the borrower to repay. Certainly bankers are desirous of their borrowers making a profit so that they will be able to stay in business and continue borrowing money (paying interest) . . . but their primary concern, first and foremost, is only to be assured that the borrower can repay the loan. Bankers do not share in the profits. Interest charges remain the same whether the borrower makes money or loses money.

The expression "Looks good on paper", probably began with balance sheets as they can be very misleading. The biggest area of potential misrepresentation comes in placing value on the assets. The most common values used are the "Book" values. Book values are usually the actual prices paid for the assets, minus depreciation. Due to appreciation and/or depreciation in market values, book values can be a long way from the actual sale value of the asset. With equipment, due to obsolescence or unusual wear and tear, the depreciation* charged against it may be inadequate. In the case of custom built equipment, such as a feedmill or fence, much of the cost incurred may be due to installation. It would therefore have a much lower market value (unless the entire operation were sold as a unit). In the case of land, because of appreciation, the actual market value may be much higher than the original book value. For these reasons, bankers may call for a balance sheet with assets reported as appraised values, rather than book values.

While balance sheets give some idea of the financial strength of a business, they give no indication of profitability! Owner's equity gives an indication of ability to repay loans, but in no way says that loan repayment can come from operating revenues. Owner's equity simply gives an indication of what could be derived (toward repaying a loan) upon foreclosure. For an indication of profitability, a second analysis is required. This analysis is known as the Income Statement.

THE INCOME STATEMENT

The purpose of the income statement is to show the profit or loss of a business for a given time period. Income statements are prepared on a yearly basis for tax purposes, and often on a quarterly or monthly basis for use in management planning.

The purpose of the income statement is to show whatever the profit or loss was for the period covered by the statement. For that reason it is sometimes also called the Profit and Loss Statement.

There are two basic ways to prepare an income statement in agriculture: (1) using cash accounting and, (2) using accrual accounting. Normally, income statements using cash accounting are prepared only for income tax purposes (page 164). This is because adjustments are not made for inventory; e.g. if cattle are fed but not sold during the accounting period, the cost of the feed is deducted as an expense but the weight the cattle

---

*Depreciation is a periodic (usually yearly) deduction from the value of an asset which theoretically represents the decline in value of the asset.

put on is not accounted for since they have not been sold. Obviously an income statement using cash accounting does not give an accurate estimation of what income really is. For management planning income statements using accural accounting are considered more accurate, since inventory allowances are made. Even then, one must be careful in determining incomes with unsold farm commodities (inventories) included, since market prices fluctuate constantly.

Another area in income statements that can be misleading is the charging off of depreciation. Capital items which are used over a period of years must be charged off as expense over that same number of years. Ideally, the depreciation expense used should represent the true cost to the business. However, in actual practice, obtaining a true or representative value can be quite difficult. In many cases, due to price escalation, the original purchase price may be a fraction of the replacement cost of the item. Probably the most accurate method of computing depreciation costs might be by figuring the difference in market value of the asset between the start and the end of the accounting period. Using this type of method will be inconsistent with computing depreciation for income tax purposes, since the IRS demands that a mathematical formula be used. By this time the reader is probably confused. The point of the whole discussion, however, is that income statements can be misleading (confusing); e.g. fluctuating prices in farm commodities (inventories) and unrealistic depreciation charges can make a business look much better or worse than it really is. For this reason, most large corporations utilize auditing firms to examine their income and financial statements before they are presented to the stockholders. What the auditing firm does is state that the figures presented are honest and realistic. In essence, the reader has the reputation of the auditing firm as an assurance that the statement he is reading is realistic. An example income statement is shown in Table 10-2.

Learning to evaluate financial data is certainly beyond the scope of this discussion. The author would like to point out, however, that while $60,965 income or profit may seem like a relatively large sum in dollars, in comparision with the amount of money handled (over 4½ million dollars), it is a very small margin . . . a return of only 1.35% ($60,965 ÷ $4,528,342), which is very poor indeed.

It should also be noted that nearly $500,000 of inventory is tied up in cattle. A 12% reduction in the cattle market (which could easily occur) would wipe out the entire profit for the year. A 20% drop in the cattle market (which could also occur), would create a $40,000 loss.

Table 10-2.   INCOME STATEMENT EXAMPLE.

Income Statement (Jan. 1, 1979-Jan. 1, 1980)

Goinbroake Feedlot
Tuffluck, Texas

Receipts

Payment on feed bills (custom feeding)	$1,825,013.06
Sale of company cattle	2,012,850.11
Accounts receivable	24,410.09
Total receipts	$3,862,273.26

Adjustments to Inventory

Cattle (840 hd @ $595 hd (850 lb @ $.70/lb)	499,800
Feed	
grain (345 tons @ $86/ton)	29,670
silage (9,250 tons @ $21/ton)	194,250
supplement (15 tons @ $221/ton)	3,315
Total Inventory	$727,035

Expenses

Feed purchases	$2,254,161.30
Cattle purchase costs	1,916,250.00
Labor	74,318.00
Interest	194,000.00
Depreciation expense	11,000.00
Vet. medicines, supplies & services	53,504.12
Bad debts	5,114.04
Office supplies & postage	4,812.16
Consulting fees	6,000.00
Miscellaneous expense	9,183.01
Total expenses	$4,528,342.61

Total Receipts plus Inventory		$4,589,308.26
Less Expenses	$4,428,342.61	
Net Income		$60,965.65

Something else the reader might want to look at is the amount of depreciation charged off; in this example, $11,000.   Looking back at the balance sheet, fixed assets were reported to be $236,020.33. Subtracting the land gives a remainder of $212,020.33 of depreciable assets. The amount of depreciation charged against those assets ($11,000), amounts to only about 5.2%, which might be unrealistically low, especially since the feed mill and trucks make up the majority of the depreciable assets . . . items that are subject to considerable wear and tear. No indication is given as to how depreciation was computed, and so this may be something the reader would want to check into further.

Again, the purpose of this section has in no way been intended to teach financial analysis as that is a discipline all its own and far beyond the scope of this text. Rather, the purpose has been to simply explain some of the variables involved in making up financial reports, and to make the point that the reader should look farther than just the bottom line.

THE CASH FLOW ANALYSIS

The purpose of the cash flow analysis is to lay out the month by month, week by week, or even day by day flow of funds to and from the business. While usually done as a projection, the purpose is to insure that the business will always have enough cash on hand to meet its obligations.

More and more bankers are demanding to see projected cash flow analyses before lending since it will point out directly whether the loan can be repaid out of normal operating funds. The balance sheet may show the ability to pay, but not whether payment can come out of operating funds; i.e. all it shows is if payment could be made on a foreclosure (if the business assets are used as collateral).

Regardless of whether bankers want to see cash flow projections or not, it is a good business practice to prepare them. This is because it is entirely possible for businesses of substantial financial strength to get in the position of not being able to meet their short term financial obligations . . . and when that happens, it can have grave consequences for the firm. In essence, it is entirely possible for a profitable business to go bankrupt. If the reader gets nothing more out of this section, it should be that profit is not a guarantee of success . . . that even profitable businesses can go broke if they cannot meet short term budgets.

Ranches typically have the greatest cash flow problem since money is expensed all year long, but income generally occurs only once a year (when their cattle are sold). For this reason, the author has chosen a ranch example to demonstrate the need for cash flow budgeting. Table 10-3 is a projected income statement for a ranch with a reasonably bright profit picture.

Table 10-4 represents a cash flow projection for the same ranch whose net income is reported in Table 10-3. As can be seen, preparing a cash flow budget is a simple matter of putting down

Table 10-3.   PROJECTED INCOME STATEMENT EXAMPLE.

Rexall Ranger Ranch
No Pesos, New Mexico
Projected Income Statement 1979

Receipts

Sale of Calves	
128 steers, 51,200 lb @ $.82/lb	$41.984
102 heifers, 38,760 lb @ $.74/lb	28,682
Sale of cull cows and bulls	
26 cows, 23,400 lb @ $.52/lb	$12,168
2 bulls, 2,250 lb @ $.62/lb	1,395
Total Receipts	$84,229

Expenses

Hired labor	$9,600
Machinery repair (windmills & pickup trucks)	650
Payments to custom hay operator ($25/ton x 320 tons)	8,000
Purchase of bulls (2)	1,800
Protein & mineral supplements	2,920
Fence repairs	3,500
Vet expense	1,500
Property taxes	1,650
Interest on working capital	1,425
Interest on mortgage	12,375
Miscellaneous	1,200
Principle payment on land	10,000
Total expenses	$41,325
Net Income	$29,609

when cash will be received and spent.   The great importance of it can be seen in the months of July, August, and September. Even though the ranch is projected to have a net profit of over

Rexall Ranger Ranch
Cash Flow

Receipts	Jan.	Feb.	March	April	May	June
Operating capital loan	15,000					
Sale of steer calves						
Sale of heifer calves						
Sale of cull cows						
Sale of cull bulls						
Total	15,000					
Expenses						
Hired labor	800	800	800	800	800	800
Machinery repair			250	100		
Custom hay paymt.						
Bull purchase						1,800
Protein & Min. supplement	600	800				620
Fence repairs			1,000	1,500	1,000	
Vet expense			300	300	200	200
Property taxes						825
Interest, working capital						
Interest, mortgage						
Miscellaneous	100	100	100	100	100	100
Operating capital						
Principle paymt. on land						
Total	1,500	1,700	2,450	2,800	2,100	4,345
Balance	13,500	11,800	9,350	6,550	4,450	105

Table 10-4. EXAMPLE CASH FLOW PROJECTION.

July	Aug.	Sept.	Oct.	Nov.	Dec.	Total for year
						$15,000
			41,984			$41,984
			28,682			$28,682
			12,168			$12,168
		1,395				$1,395
		1,395	82,834			$99,229
800	800	800	800	800	800	$9,600
					300	$650
4,000	4,000					$8,000
						$1,800
				400	500	$2,920
						$3,500
200			300			$1,500
					825	$1,650
					1,425	$1,425
					12,375	$12,375
100	100	100	100	100	100	$1,200
					15,000	$15,000
					10,000	$10,000
5,100	4,900	900	1,200	1,300	41,325	$69,620
(-4,995)	(-9,895)	(-9,400)	72,234	70,934	29,609	$29,609

$29,000* . . . during the months of July, August, and September the ranch will have a deficit of funds. Knowing about the deficit in advance (through the use of the cash flow projection), allows time for arranging in advance for a short term loan.

The danger in not setting up a cash flow projection, of course, is coming up on the deficit months unexpectantly, and having to obtain emergency loans. If emergency loans cannot be obtained, the situation can become serious. In this example the rancher would be unable to pay the custom hay operator for his services. If the hay operator refused to work, and the hay was needed for winter feed, it could have obviously grave consequences for the ranch.

A similar situation exists with feedlots that buy silage as their main source of roughage. Silage is only cut once a year, and so the entire year's supply must be purchased within what is usually no more than a 30 day period. For feedlot owners with extensive incomes, this is a boon since it allows the expensing of extremely large sums of money near the end of the year. But for feedlots of modest means, it creates a serious cash flow problem. Silage is only available at harvest, and unless the feedlot obtains its yearly needs at that time, it will be forced into purchasing other roughages later on in the year which are typically more expensive (in most areas silage is usually the cheapest available roughage on an energy basis).

SUMMARY

The purpose of this chapter has not been to instill any significant degree of expertise in the subjects discussed. Indeed, each of the subjects discussed are quite complex and would require an entire text to cover them thoroughly. Accordingly, the purpose of this section has been only to point out the significance of the topics discussed, and the need for obtaining either more knowledge or professional consultation when dealing with the subjects.

---

*For clarity of explanation, the net cash flow figure and the net income from the Income Statement have been shown to be an equal amount ($29,609). In a real situation the figures would actually be different since Income Statements would include a depreciation expense (decline in value of capital assets). Depreciation expenses are not included in cash flow projections, since funds do not actually flow out (for depreciation).

CHAPTER 11   BEEF CATTLE NUTRITION

RUMINANT DIGESTION

Cattle are ruminants and therefore a basic understanding of ruminant digestion is required in order to fully understand beef cattle nutrition.   Most texts refer to ruminants as animals having stomachs with four compartments.   While ruminants do have four separate compartments, it is more important to recognize that ruminants have two separate digestive systems; i.e. there are four stomach compartments, but together they comprise two different types of digestive systems.

Anatomically the four compartments may be loosely referred to as stomachs:

1. rumen
2. reticulum
3. omasum
4. abomasum

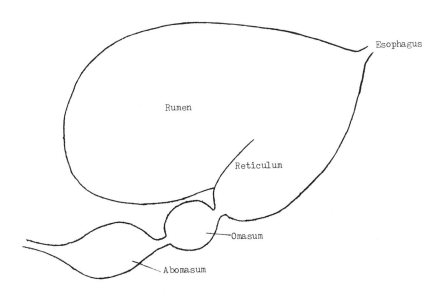

Figure 11-1. RUMINANT DIGESTIVE TRACT.

For all practical purposes, only two of the compartments need be remembered for a basic understanding of the digestion patterns in ruminants. Those two compartments are the rumen and the abomasum. Essentially the reticulum is part of the rumen . . . . the two organs are often referred to as the reticulo-rumen, and the function of the omasum is not clearly understood. (The only reason for mentioning all four compartments was so the reader would not be confused when only two digestive systems are mentioned - when he has heard that cattle have four stomachs.) The author would prefer that the reader picture the digestive tract of ruminants looking like what is represented in Figure 11-2; i.e. concentrate only on the rumen and abomasum:

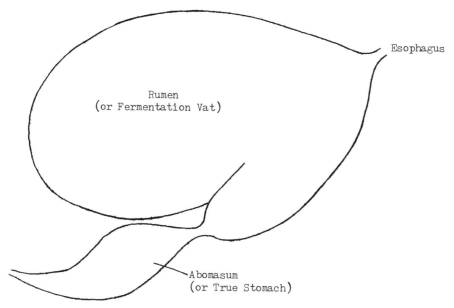

Figure 11-2. RUMEN AND TRUE STOMACH (ABOMASUM).

Together the rumen and the abomasum represent two completely different digestion systems. It was mentioned that the compartments can only be loosely referred to as stomachs. This is because the abomasum is the only compartment which is by definition a true stomach . . . hence the abomasum is often referred to as "the true stomach". The abomasum contains and secretes much the same type enzymes and acids found in the stomachs of monogastric animals such as the pig. Feeds are therefore digested in the abomasum in the same manner as in monogastric animals.

The rumen, on the other hand, is not a true stomach. It neither contains nor secretes any digestive juices. The only similarity between the rumen and a true stomach is that it holds feed. The rumen is actually nothing more than a large fermentation vat.

Normally the rumen is filled to about ⅔ to ¾ capacity with fluid and ingesta. Contained in the rumen fluid are countless billions of microorganisms which are actually responsible for the digestion that takes place in the rumen.

These microorganisms have the ability to break down cellulose, which is the main form of energy in fibrous feeds. Simple stomached animals do not have the enzymes required to break down cellulose or fiber. As a result then, most simple stomached animals cannot adequately utilize high fiber, low quality feeds;* i.e. grasses, hays, silages, vegetable by-products, etc. (feeds commonly referred to as roughages). Because of the microorganisms then, ruminants have the ability to digest fibrous feeds.

RUMINANT PROTEIN DIGESTION

In ruminant protein digestion, rumen microorganisms break down the ingested proteins for their own use. Proteins, of course, are made up of amino acids which simply described, are organic molecules with a nitrogen atom attached. The nitrogen is indispensable, for without nitrogen, there is no protein.

The rumen microorganisms break the nitrogen away from the rest of the protein molecule, and resynthesize it into protein for their own use. They then use the protein to reproduce by division. Eventually the microorganisms are moved from the rumen into the abomasum. They are then digested by the digestive juices in the abomasum or the small intestine and used as a source of protein themselves (see Figure 11-3). In essence, whenever a ruminant eats a proteinaceous feedstuff, it is actually feeding the microorganisms contained within the rumen, which are in turn used as a feedstuff themselves.

Since the rumen microorganisms break down most of the protein the animal eats, and then resynthesize it, the amino acid balance is of little concern under most circumstances. This is why non-protein nitrogen compounds such as urea may be used for part of the required protein (see Chapter 12). In essence ruminants have a nitrogen requirement, not a protein requirement per se.** With monogastric animals the amino acid content of the proteins eaten must be balanced, or the animal can't effectively utilize it.

---

*Exceptions would be horses, rabbits, etc., which have developed specialized sections in the large intestine which function as fermentation vats.

**In certain special situations there are circumstances where amino acid balance is of some importance. These situations are discussed separately on page 57, and page 104. Also, amino acid balance is of particular importance in very young calves.

Figure 11-3. RUMINANT PROTEIN DIGESTION.

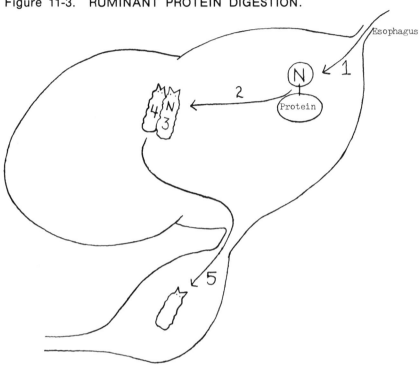

1. Animal eats protein.
2. Microorganism breaks down the protein and removes the nitrogen.
3. Microorganism uses the nitrogen to synthesize protein for its own use.
4. Microorganism uses the protein to reproduce by division.
5. Microorganism moves to the abomasum where digestive juices break it down and use the proteins it has formed, for the use of the host animal.

ENERGY DIGESTION IN THE RUMEN

Starch and cellulose are the primary forms of energy digested in the rumen. Starch is the primary form of energy found in grains, and cellulose the primary form of energy found in roughages. Both starch and cellulose are comprised of numerous units of glucose hooked together with two distinctly different types of bonding. Starch is hooked together with what is known as alpha bonding, and cellulose is hooked together with what is known as beta bonding.

The chemical dynamics involved in the two forms of bonding are beyond the scope of this text. All the reader need understand is that alpha bonds are broken quite easily, but beta bonds are extremely difficult to break. Only microorganisms have the enzymes required to break beta bonds, and therefore only ruminants and a few other selected animals with fermentative digestion (horses, rabbits, etc.) can utilize cellulose as a major energy source.

Once the bonds are broken, be it starch or cellulose, the main energy product released is glucose. The actual chemical structure of glucose is represented by Figure 11-4. (If the reader does not have extensive training in chemistry please read the footnote below.)*

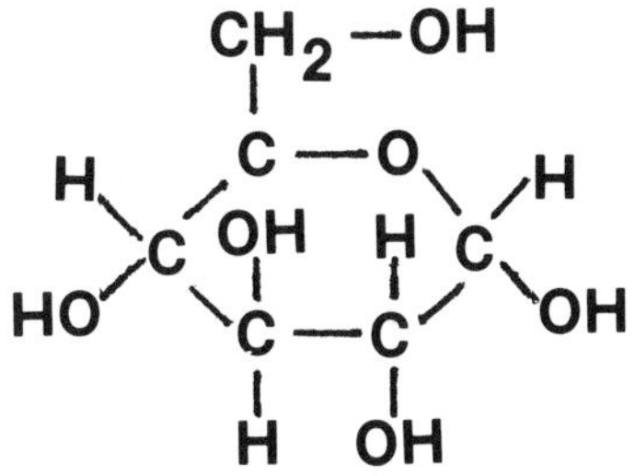

$$CH_2-OH$$

Figure 11-4. GLUCOSE MOLECULE.

For the discussion to follow, however, the reader can forget about the hydrogen (H) and hydroxyl groups (OH). For a basic understanding of ruminant energy digestion all one need recognize about the glucose molecule is that it contains 6 carbons and 1 oxygen hooked together in a ring:

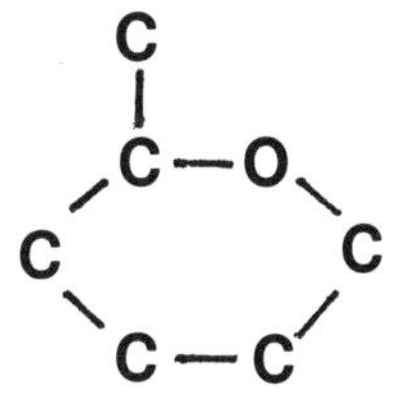

Figure 11-5. GLUCOSE CARBON RING.

*For those who have limited formal training in chemistry, please do not be "turned off" by the chemical structures. If the structures themselves to not mean anything to you, please do not be concerned, as they are not intended to. Simply look as the structures and follow the descriptions in the text, and an understanding of their purpose should become clear. If something is confusing, go ahead and read all the way through the section. At that point, an overall understanding should become evident and therefore individual points should become more clear. While the inclusion of chemical structures may seem to make this a very advanced discussion, the actual purpose has been to make the basics more clear. Moreover, if the reader makes the extra effort to understand the purpose of the structures his understanding of many different aspects of beef production will be greatly enhanced.

Hooked together, the glucose units form starch, or cellulose, depending upon the type of bonding:

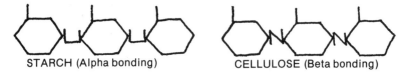

STARCH (Alpha bonding)   CELLULOSE (Beta bonding)

Figure 11-6. STARCH AND CELLULOSE BONDING.

Once the glucose molecule is released, the rumen microorganisms then break the individual glucose units into 3 carbon units known as pyruvic acid.

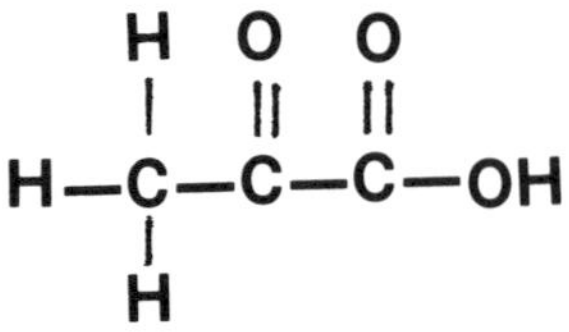

$$H-\underset{\underset{H}{|}}{\overset{\overset{H}{|}}{C}}-\overset{\overset{O}{\|}}{C}-\overset{\overset{O}{\|}}{C}-OH$$

Figure 11-7. PYRUVIC ACID.

As explained about the glucose molecule, the only thing the reader need recognize is the carbon chain (C-C-C). In a diagram, the situation would look like this:

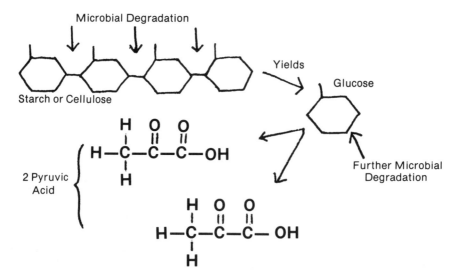

Figure 11-8.   MICROBIAL DIGESTION OF STARCH OR CELLULOSE.

Pyruvic acid is then subjected to further microbial action to form what are known as the Volatile Fatty Acids (VFAs), sometimes known as the Short Chain Fatty Acids. There are several different VFAs but the most important are acetic and propionic:

ACETIC ACID   PROPIONIC ACID

Figure 11-9. VOLATILE FATTY ACIDS.

The VFAs are absorbed through the rumen wall and used for energy. VFAs comprise the ruminant's major source of energy.

Acetic acid is the VFA found in the greatest concentration. When the 3 carbon pyruvic acid molecule is broken down to form the 2 carbon acetic acid, a carbon atom is lost.

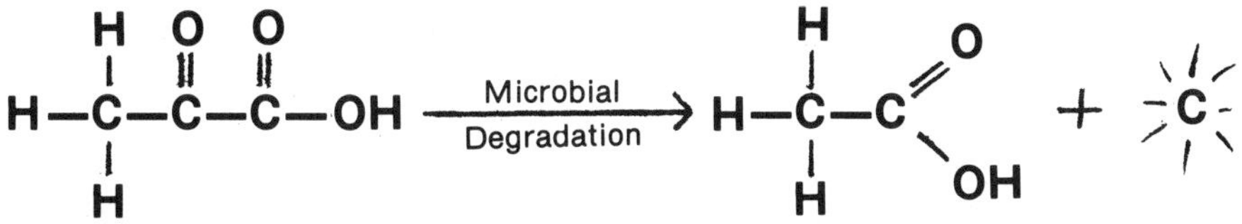

PYRUVIC ACID   ACETIC ACID

Figure 11-10. ANAEROBIC GLYCOLYSIS (EMBDEN - MYERHOFF PATHWAY).

The carbon released in the formation of acetic acid from pyruvic acid goes to form methane (CH_4) gas. This is the gas that cattle must eructate (belch) off continually - or if for some reason they cannot, then this is the gas that causes them to bloat.

Bloat, of course, causes many cattle to die, and is a fairly important loss to the cattle industry. But almost immeasurably more important as a loss to the cattle industry . . . is the loss of the gas itself, and the chemical reactions that cause it to be lost.

Carbon is the basis of all energy,* and therefore the loss of

*Glucose and VFAs are the form of carbon used by animals for energy. Other forms of energy such as propane gas (H-C-C-C-H) and butane (H-C-C-C-C-H) are also carbon chains. Gasoline is also comprised of carbon chains; octane, simply being an eight carbon chain compound.

carbon in the form of gas is a serious energy loss. Obviously if the animal belches this gas (CH_4) off, it cannot use the carbon contained in it for energy.

An even greater loss is incurred by the energy that must be used by the microorganisms to conduct the reactions that liberate the gas. This is known as the Heat Increment or Fermentative Heat Loss. The bonds that hold cellulose together are particularly strong, and so considerably more energy is required to break the bonds of cellulose (as compared to starch). However, the loss from the fermentation of either substance is substantial.

The combined energy loss from gaseous and fermentative energy losses will vary with the type of feed used. As a general rule, it may be said that approximately ⅓ of all energy digested in the rumen is lost. This energy loss is the primary reason that ruminants are poor converters of grain; the primary reason why hogs and other monogastric animals are so much more efficient; i.e. feedlot grain ration conversion to meat with beef cattle will run between 7 and 8 to 1, whereas hogs will convert at a ratio of 3.5 to 4:1. The hog of course, does not have a rumen, and so the starch from the grain it eats is digested and absorbed as glucose, so there are essentially no fermentative energy losses.

But while the ruminant is not as efficient as monogastric animals at converting grain to meat, it is able to digest fibrous materials which the monogastric animal cannot utilize. This, of course, is why the beef industry will always be a viable industry - because there will always be fibrous materials around that could not be used otherwise. Then too, as pointed out in the introduction, it should be remembered that in terms of total conversion of grain to meat, the feedlot steer is just as efficient as the hog, because only 30 to 40% of the animal's final weight is attributable to grain rations.

RUMINANT USE OF FAT

Most feedstuffs fed to beef cattle do not contain appreciable amounts of oil or fats. In times past most oil meals (cottonseed meal, soybean meal, etc.) contained substantial amounts of oil. Most oil mills now use the solvent extraction process, and so most of the oil is removed. Probably the only major feedstuff available with a high oil content is whole cottonseed, and most of it is used for the feeding of dairy cows.

Beef cattle can, however, utilize small quantities of fat, and are often fed tallow or feed grade fat in feedlot rations. The textbook energy value for fat is given as 2.25 times the energy value of carbohydrates. For use in grain rations for cattle, the energy value of fat vs. carbohydrates will be somewhat greater than that.

The actual mechanism is not clearly understood, but fat is known to inhibit the production of methane (CH_4) gas. As pointed out in the section on energy digestion, methane gas in an energy loss, and so the decrease in production of methane is an energy saving. Fat will therefore show a gain response (on grain rations) in excess of what can be attributed to the energy it contributes to the ration. However, at levels higher than 4% of the ration, fat will usually reduce intake, and can cause cattle to go off feed.

In high roughage rations fat apparently depresses the utilization of fiber. Again, the actual mechanism is not understood, except that cellulose digesting bacteria are somehow inhibited by the presence of fat.

UTILIZATION OF LOW QUALITY ROUGHAGES

The fiber content of roughages limits the amount of roughage a ruminant can consume. The higher the fiber content the lower the intake. This is primarily due to two factors:

(1.) Cellulose, the most common carbohydrate in roughages is difficult for the rumen microorganisms to break down. As a result, cellulose is broken down much more slowly than other carbohydrates. Since it is broken down relatively slowly, then passage of cellulose through the rumen is also slow.

(2.) Lignin, a structural component in fiber, is almost completely indigestible. Even the rumen microorganisms cannot break down lignin to any degree. As the fiber content of the roughage goes up, so does the lignin content.

Cellulose and lignin therefore accumulate in the rumen. This causes the animal to feel full, which eventually reduces intake.

As the fiber content of roughages increases, the protein level usually decreases. Particularly serious is the fact that as the total protein decreases, the amount of digestible protein decreases more rapidly. In really low quality roughages, not only is there not enough protein to meet the needs of the animal, but often there is not even enough protein to maintain an active rumen microbial population.

From 3 to 5% of the nitrogen contained in roughage is complexed with the lignin, or indigestible fraction of the fiber. Straw and cornstalks will show laboratory protein values of 3 to 5%, and many grass hays will have only 5 to 7% protein. In essence, straw and cornstalks will have 0 to 2% digestible protein. Under most circumstances that is not enough protein to maintain a healthy population of rumen microorganisms. As a result then, the rumen microorganism population decreases.

As the microorganism population decreases, the rate of digestion decreases at a greater rate. Microorganisms are responsible for breaking down roughages, and so as the numbers of microorganisms decrease, so does the rate of roughage digestion. "The less bugs there are to work on the fiber, the slower it is going to be utilized". The slower it is utilized, the more the fiber builds up in the rumen, and the fuller the animal feels. Intake therefore decreases rapidly after the first few days. For example, hungry cattle can eat up to 3% of their bodyweight in poor quality grass or hay for two or three days. Over a longer period, however, intake would be closer to 1.25-1.75% of their bodyweight.

PROTEIN SUPPLEMENTATION FOR CATTLE FED LOW QUALITY ROUGHAGES

Protein supplementation is necessary for cattle that are going to be maintained on low quality roughages such as straw or cornstalks. The purpose is not so much to meet the animal's protein requirement, as it is to make protein available to the rumen microorganisms. Without proper protein supplementation the rumen microorganism population cannot reproduce properly, which results in reduced passage of ingesta through the digestive tract. As explained earlier, this makes the animal feel full and thereby reduces intake.

Through proper protein supplementation, the rumen microorganisms are able to multiply which in turn increases the rate of passage through the digestive tract; i.e. with a larger population of cellulose digesters (the microorganisms) in the rumen, the faster the cellulose will be digested, and thus the cattle can eat more of the roughage (see also page 29).

TERMS USED IN BEEF CATTLE NUTRITION

DRY MATTER (DM)

Dry matter is the weight fraction of the feed left after all the moisture is driven off. Hay and grains will normally vary between 85 and 92% dry matter. Silages will usually run between 28 and 40% dry matter.

CRUDE PROTEIN (CP)

Crude protein is an expression of protein obtained by laboratory analysis. The percent nitrogen in a feed is analyzed, and then multiplied times 6.25 since nitrogen is contained in most proteins at a rate of 16% by weight (the reciprocal of 6.25). The term gives no indication of the actual digestibility or quality of the protein, hence the name crude protein.

NON-PROTEIN NITROGEN (NPN)

As pointed out in the section on ruminant protein digestion, the rumen microorganisms do not have a protein requirement so much as they have a N requirement. For this reason, compounds that contain N, but are not protein, can be used to provide a portion of the ruminant's protein requirement. Urea is the most common NPN compound used commercially, although various ammonium compounds are also used.

The amount of N contained in these compounds by weight is usually quite high. They therefore have rather high crude protein ratings; e.g. urea has a crude protein equivalency of 281%. The cost of these compounds per unit of nitrogen is usually much lower than natural protein sources. There are a number of factors that affect how much of these compounds may be used. Often these factors are overlooked, and so in some circles NPN is a controversial subject. For this reason, a separate chapter has been written to describe the proper use of urea (Chapter 12).

DIGESTIBLE PROTEIN (DP)

Digestible protein is a measure of the amount of natural protein that can actually be digested and absorbed by an animal. It is determined through animal metabolism trials, and is useful for comparing different kinds of feeds.

Since DP is determined with animal trials, it cannot be used on a practical basis for analyzing an individual feed; i.e. it normally requires a 6 week feeding trial in special metabolism stalls, involving daily feces and urine collection and a substantial amount of lab work.

PROTEIN SOLUBILITY

Protein solubility is a term that has come into fairly common usage lately. A controversial topic, protein solubility refers to the measure of the rapidity by which a protein will go into solution in rumen fluid. A protein that goes into solution rather quickly will usually, but not always, be more vulnerable to microbial attack.

CRUDE FIBER (CF)

Crude fiber is a laboratory analysis for estimating the amount of fiber in feedstuffs. Since fiber is not as digestible as other forms of carbohydrates, something about the quality of a feedstuff is reflected by the fiber content; generally, the higher the fiber content, the lower the digestibility of the feed.

ACID DETERGENT FIBER (ADF)

This is a refinement to the crude fiber analysis developed by Van Soest. This method attempts to more closely indentify the material that is less digestible to the animal.

ADF analyses are usually higher (percentage wise) than CF analyses because it includes some attached proteins and some hemicellulose. For this reason, the two are not directly comparable. This can cause confusion since many laboratories are now reporting fiber as ADF, whereas most feed texts report fiber as CF. A conversion factor of .89 is satisfactory for converting ADF to CF for most grass-like forages (Van Soest, 1976).*

ASH

As the name implies, ash is what is left over after a feed sample is burned. It has often been said that the analysis is of little value since individual minerals are not identified. For the most part that is true, but there are practical applications for the Ash analysis. For instance, a high ash report on a grain sample would indicate considerable dirt in the fines (dust and broken kernel part of the sample). A high ash content in a grass sample could indicate a high silica content (high silica can cause urinary calculi in grazing steers).

TDN (TOTAL DIGESTIBLE NUTRIENTS)

Total Digestible Nutrients, commonly spoken of as TDN, is used as a measure of the energy contained in a feed. It is actually a measure of the digestibility of all the organic constituents of a feedstuff (which would include protein), but since the digestibility and energy value are so closely related, the term is considered synonymous with energy.

NET ENERGY

Net energy is a system of measuring the energy content of feeds which is available to the animal for some productive function. To arrive at a value, animals are given the feed in question under carefully monitored environmental conditions.

As mentioned earlier, there are large gas and heat losses in rumen fermentation. In net energy experiments these losses are measured; i.e. the amount of CO_2 and CH_4 given off by the animal are measured, along with the heat radiated from the animal's body. This, when combined with losses in the urine and

*Personal communication.

feces, yields a value that represents only the energy that can actually be utilized for production.

In practical application, net energy has its greatest advantage over the TDN system in evaluating roughages. TDN is only a measure of what an animal takes in, versus what is left in the feces. There is no allowance made for the gas and heat loss. Since roughages contain much more fiber than concentrates, much more fermentation is required to digest them, and therefore there are much larger gas and heat losses with roughages than with concentrates. The TDN system therefore overestimates the energy value for roughages.

For example, the net energy and TDN systems rate the following feedstuffs accordingly (Morrison, 1961):

Feedstuff	TDN[1]	NE[1]	% Difference
Corn	94.1	94.23	0.1%
Milo	89.21	87.41	2.0%
Alfalfa hay	56.02	44.86	19.0%
Wheat straw	43.84	10.79	75.4%

[1] Dry matter basis.

As would be expected, there is very little difference between the TDN and net energy systems for the grains, but note the wide variation for the roughages. Note also the great difference within roughages.

The TDN system rates wheat straw as having 78% the energy of alfalfa hay. That is certainly not true as cattle fed wheat straw will usually lose weight, whereas cattle fed alfalfa hay can gain up to 1.0 lb/day. The Net Energy System rates wheat straw as having only 24% the energy of alfalfa hay, which is much more nearly correct.

CALIFORNIA NET ENERGY SYSTEM

This is a refinement to the Net Energy System developed by California researchers Lofgreen and Garret. It is designed for predicting performance on feedlot cattle fed high grain rations and implanted with growth hormones. For that purpose, it does appear to be somewhat more accurate than the old Net Energy or TDN systems.

Under the old systems, a single value is given as the estimate of energy in the feed. With the California system, separate estimates are given for energy for maintenance (NEmain), and energy for gain (NEgain) (see also page 133).

GUARANTEED ANALYSIS

Most states and provinces have laws which require that manufactured feeds carry a guaranteed analysis. For this purpose a series of laboratory analyses are listed on the guarantee, which will usually consist of:

Crude protein (CP)
Non-protein Nitrogen (NPN)- as a percent of CP
Phosphorous
Calcium (maximum)
Calcium (minimum)
Crude Fiber
Crude Fat

In addition to this, any medication or other drug must be specified, and instructions for its feeding, and withdrawal must be included. Oftentimes the manufacturer will guarantee analysis for vitamins (usually vit. A), and sometimes minerals other than Ca and P.

It is important to realize that the guaranteed analysis is a laboratory analysis only. The values listed represent only chemical findings, and tell nothing of the digestibility of the feed.

MINERALS

A WORD ABOUT MINERAL NUTRITION

Minerals, of course, play an extremely important role in beef cattle nutrition. Gross evidence of this is exemplified by the many deficiency symptoms that manifest themselves when certain minerals are not present in adequate amounts. Except for a few isolated areas, however, clinical symptoms of deficiencies do not ordinarily occur.

The problem is that in most instances the deficiencies that do occur are usually subclinical in nature; i.e. the animals appear to be perfectly normal and healthy, but less than optimum amounts of minerals are being supplied, and so the animals are not as productive as they should or could be. Adding to the problem is the fact that requirements for most minerals are not precisely known for all situations.

Probably the biggest reason that precise requirements are not known in many instances, is due to the many complex interrelationships between minerals. That is, most minerals have many physiological functions, and there are usually other minerals that can substitute for at least some of their functions. Also excesses of some minerals can tie up or prevent utilization of other minerals.

TYPES OF MINERALS

Minerals are normally classified into two categories; (1.) macro minerals and (2.) micro or trace minerals. Macro minerals are those which are required in relatively large amounts, such as sodium (Na), calcium (Ca), phosphorous (P), potassium (K), and magnesium (Mg). The micro minerals are required in very small amounts and are expressed in parts per million, whereas macro minerals are usually expressed in terms of ration percent (e.g. .3%), or grams (e.g. 18 grams). Examples of trace minerals that beef cattle are known to have a need for are; cobalt (Co), copper (Cu), zinc (Zn), manganese (Mn), iron (Fe), iodine (I), and selenium (Se).

INDIVIDUAL MACRO MINERALS

CALCIUM

Calcium is the largest constituent of bones and is found in greater quantities in the body than any other mineral. In addition to being used in the formation of bones and teeth, Ca has some absolutely vital physiological roles in the soft tissues.

While the exact process is not completely understood, Ca is known to be used in muscle contraction. Without adequate Ca the animal will go into a state of tetany and eventually will die if not treated. This situation does not occur in beef cattle as a simple deficiency, but is a fairly common occurrence in dairy cattle. Known as milk fever, it is a deficiency of Ca created by the onset of lactation; i.e. when high producing dairy cows freshen, oftentimes most of the available Ca will be put in the milk, thereby leaving an inadequate amount for the cow's bodily needs.

Deficiency Symptoms - The classic deficiency symptoms of Ca is the condition known as rickets. Rickets occurs as a deficiency of Ca during the early growth stages of a young animal. The condition results in a buckling of the knee joints, giving the animal a bow-legged look.

In adult animals a Ca deficiency results in a condition known as osteomalacia. Calcium (as well as P and Mg) is stored in the bone. It can be pulled out of the bone as required, and used for physiological purposes. In the case of a deficiency (in an adult animal), Ca can be pulled out to the point of severely weakening the bone. As a result, stresses that would ordinarily cause no injury, will result in bone fractures.

As a practical matter, rickets and osteomalacia caused by a Ca deficiency do not occur in beef cattle, as most forages contain considerable amounts of Ca. Clinical symptoms of rickets and osteomalacia do occur however, but they are almost always caused by a P deficiency.

In order for a clinical case of Ca rickets to occur, a calf would have to be weaned off its mother at an early age and fed low quality roughage, or given a concentrate ration without supplemental Ca. For a clinical case of osteomalacia to occur, cattle would have to be held on a grain ration without supplemental Ca.

PHOSPHOROUS

The discovery of P as a required nutrient for beef cattle was probably the greatest single breakthrough in the history of beef cattle science. Phosphorous is, of course, used in teeth and bone formation, but probably more important from a production point of view . . . P is an essential nutrient for reproduction. Phosphorous supplementation typically increases calf crop (at weaning) from 5 to 15%. In seriously deficient areas, P supplementation can increase calf crop as much as 40%.

Phosphorous is also a vital element in energy metabolism. As the energy or grain content of a ration goes up, so does the requirement for P. Without P, energy cannot be utilized. For example, for a 660 lb steer to maintain his bodyweight, he needs about 8.1 grams of P (Appendix Table II). But to gain 1.1 lb, he needs 13.8 grams of P, more than a 60% increase.

Deficiency Symptoms - As mentioned in the Ca section, some of the deficiency symptoms of P are the same as Ca; i.e. rickets and osteomalacia. While P is required for energy utilization, tetany will not occur as P will be pulled out of the bones - hence the osteomalacia. Cattle will, however, become rather weak and appear debilitated.

Rickets and osteomalacia as caused by P, do occur with some frequency. This is because most forages are relatively low in P and range animals must either be supplemented with P, or fed concentrates (which contain relatively high levels of P). Milk is of course, high in P, but rickets can occur in suckling calves as a P deficiency can greatly reduce milk production. However, the P deficiency must be quite severe and prolonged for milk production to decrease to the point of causing a clinical case of rickets (in the calf). More often, the cow will suffer osteomalacia, and the calf's growth will merely be retarded.

On a herd basis, one might call reduced reproductive efficiency a clinical symptom of P deficiency*. Clinical symptoms are usually defined as symptoms which can be seen, and certainly a large number of barren cows is something that can be seen. As the situation becomes more aggravated, osteomalacia will develop.

One of the first signs of P deficiency is what is known as pica or depraved appetite. The animal will instinctively chew on foreign objects such as wood, bones, etc., in an attempt to obtain P. (In a feedlot/high grain situation, chewing on wooden fences, etc. is more of an indication of K deficiency - eating dirt can be an indication of a deficiency of a number of minerals, including salt.) Chewing on bones can cause serious secondary bacterial diseases caused by clostridial organisms found in the decaying bone marrow.

Figure 11-11. The eating of dirt is a common feedlot problem. In some cases it may be an indication of a mineral deficiency, and in some cases is just a product of boredom.

CALCIUM:PHOSPHOROUS RATIOS

Excesses of either Ca or P can tie up and prevent utilization of the other mineral, and so Ca:P ratios can become relatively

*One of the most comprehensive studies was done by the Texas Agricultural Experiment Station (Reynolds, 1953). Over a 4 year period cows supplemented with P weaned a 92% calf crop and had a 365.3-day calving interval, whereas unsupplemented cows had a 64% calf crop and a 459-day calving interval.

important. In most species of animals, Ca:P ratios do become somewhat critical. With beef cattle a somewhat wider range of ratios can be tolerated, with one exception (discussed later in this section).

Calcium and P are found in bones at a ratio of 2:1, and for many years 2:1 was considered the ideal ratio at which to feed Ca and P. Phosphorous was later found to be required in energy metabolism, and so as animals became more productive (gain faster; give more milk), it was found that the amount of P required was much more than originally deemed necessary. As it is today, published Ca and P requirements for beef cattle run at ratios from 1.5:1 to 1:1.

In other words, Ca and P requirements as published by the National Research Council are on a gram or percentage basis, but when compared on a ratio basis, it works out to 1.5:1 and 1:1. Thus, with one important exception it is the total amounts rather than the ratios that one should ordinarily be concerned with.

Calcium/phosphorous ratios become extremely important with feedlot rations consisting mainly of grain sorghum (milo). A ratio of at least 1.5:1 to 2:1 must be maintained. If it is allowed to drop below 1.5:1, clinical symptoms of urinary calculi (water-belly) in steers will usually occur within 60 days.

Urinary calculi are stones made up of P, Mg, and Ca which precipitate out of the urine. Normally the stones cause no problem for heifers or bulls, but since castration decreases the size of the urethra, the stones can cause serious problems for steers. For some unknown reason the problem occurs primarily in rations with sorghum as the grain source. Extremely wide ratios of 4:1 or more have also been reported to cause urinary calculi with sorghum (Crookshank, 1978*).

MAGNESIUM

Magnesium is the third most abundant mineral found in the body. Along with Ca and P, Mg is a constituent of bone and can be mobilized from the bone. Although the mechanisms are not completely understood Mg is known to be involved in neuromuscular control. Apparently there are a number of inter-relationships between Mg and Ca. It has also been reported that K and P can reduce Mg absorption.

---

*Personal communication. (H.R. Crookshank, Dept. Veterinary Research, USDA. College Station, Texas.)

Deficiency Symptoms - The most well known deficiency symptom is what is known as "grass tetany", in which the animal goes into an acute state of tetany. While commonly known as a Mg deficiency, there are other factors involved. However, supplementation with Mg will prevent the condition.

Uncomplicated Mg deficiencies result in a condition similar to polioencephalomalacia. The animal assumes a posture with the head pulled back in what is commonly termed a "star gazing" position. If the deficiency continues, the animal will eventually go into convulsions and die. This condition is seldom seen with beef cattle. Probably the only situation that could create a Mg deficiency would be for a calf to be raised only on milk.

SODIUM AND CHLORINE (Salt)

These two minerals will be discussed together since they are usually supplemented together as salt. Both minerals have a number of functions, but probably the most prominant function is the regulation of body pH and osmotic pressure. They are also involved in neuromuscular control. Chlorine is used in HCl secreted in the abomasum, as well as a number of other digestive juices and enzymes. Sodium is a major constituent of ruminant saliva which is used as a rumen buffer.

Deficiency Symptoms - Clinical cases of NaCl deficiencies in beef cattle are quite rare. Almost every stockman realizes the necessity for salt, and since it is very inexpensive, it is usually provided. But even if salt is not provided, it has been repeatedly shown that K can substitute for Na to some extent. Most forages are quite rich in K, and so cattle can get by for substantial periods of time without supplementary Na. Chlorine is not required in nearly as large amounts as Na, and most forages contain adequate levels of Cl. This, coupled with the fact that cattle have poorly developed sweat glands, allows cattle to go for substantial periods without showing clinical symptoms of salt deficiency; i.e. sweat is the main avenue of NaCl loss when salt is deficient in the diet.

Subclinical problems, of course, can occur. As with most other minerals, the only symptoms will be poor performance, reduced weight gain, etc. But these will not likely be noticed by the cattleman, unless he has a similar group of animals with salt and compares rate of gain, weaning weight, etc.

If salt deprivation occurs over a period of time, clinical symptoms will appear. Non-specific in nature, clinical symptoms will be a very rough hair coat and gradual emaciation. Death can occur in extreme cases, but rather than go into tetany as with other minerals that aid in neuromuscular control, the animals will become extremely lethargic and eventually they just lay down and die. Apparently what happens is that the body electrolyte level becomes so low that the heart cannot continue beating; i.e. it has long been known that a solution of NaCl with lesser amounts of $CaCl_2$ and KCl are required to keep laboratory animal hearts beating, once they have been removed from the animal (Ringer's solution).

Toxicity Symptoms - Ruminants are able to absorb and excrete relatively large amounts of salt, and so acute toxicities do not often occur. When they do occur, it is almost always a result of a water restriction after taking in a large amount of salt. In the real life situation, the only time this normally occurs is when salt is added to a concentrate feed to limit its intake, and then a windmill or water pump breaks down, etc.

When more salt is present in the blood than the kidney can excrete, water will pass out of the body cells and into the blood. This mechanism is an attempt to restore the blood osmolarity back to normal. The animals will go off of feed and become lethargic and dehydrated. In some cases, death has occurred.

In many areas the water available for livestock use has high concentrations of salt. The problem this presents is that if the water becomes too salty, cattle will simply reduce their intake of water. Reduced water intake, of course, means reduced feed intake, which means performance will be reduced.

But as mentioned earlier, ruminants are able to tolerate relatively high levels of salt. Presumably this is due to the need for Na as a rumen buffer. Most experimental work has shown that cattle can tolerate salt levels in the water of up to 1% (10,000 parts per million) without any apparent ill effects. This level may be deceiving as when high levels of salt appear in the water, there are also usually high levels of Ca or Mg salts as well.

POTASSIUM

Most grasses contain about 1% K, and legumes can run as high as 2.5%; thus K is seldom a limiting mineral for range or pasture animals. Grains run about .4% K, and therefore on high concentrate rations, K is often supplemented, since the requirement runs from about .5-.7%.

As mentioned in the Na section, K can substitute for a substantial amount of Na. Specifically, K can substitute to some degree for Na in its role as an osmolarity regulator. The ability of K to substitute for Na in neuromuscular control is not as clear although it is well known that K is involved in neuromuscular control.

Deficiency Symptoms - Due to the relatively high level of K in plants, clinical deficiencies in beef cattle do not occur. Grain contains about half as much K as most forages, and so minor subclinical problems can occur with low roughage rations. In that case the only difference between K-supplemented cattle and non-supplemented cattle will be a gain response of about .1 to .2 lb in average daily gain.

Cattle that have been fed purified diets (corn starch, etc.) low in K, have developed non-specific deficiency symptoms. The cattle had an unkept, emaciated appearance. As the situation progressed, the cattle became very weak, and had a wobbling gait. In most of these experiments the cattle were reported to chew on wooden posts and fences. Pica (chewing of foreign objects) is a fairly common manifestation of mineral deficiencies, but the author has noted that even in very mild deficiencies (unsupplemented feedlot rations), cattle will instinctively begin chewing on wood (which is relatively high in K), and over a period of time can be very destructive. The addition of KCl to the ration often quels the problem.

SULFUR

Sulfur is used in a variety of body compounds; connective tissue, mucous, lipoproteins, estrogens, and other hormones. Sulfur is also contained in 2 B vitamins, biotin and thiamin; and 3 amino acids, methionine, cystine, and cysteine.

The use of urea and other NPN sources in the formulation of ruminant protein supplements has necessitated the supplementation of S in most situations. In order for rumen microorganisms to synthesize the S-containing amino acids (from NPN), they must have S available.

Deficiency Symptoms - Deficiencies of S produce non-specific symptoms that typify many different deficiency conditions. The animals will gradually go off feed, slowly emaciate, and if the condition continues, death will occur.

Toxicity Symptoms - High levels of sulfates will cause a profuse diarrhea. Sulfates are lowly digestible, and so toxicity symptoms are generally limited to diarrhea and subsequent poor performance.

If trace minerals are marginal, or sulfates are particularly high, a variety of other problems may occur. Sulfates can combine with several trace minerals, and limit their absorption. The trace mineral most often implicated is Cu, as Cu readily combines with S. Indeed, feeding sulfates has been used as a method of preventing Cu toxicity.

INDIVIDUAL TRACE MINERALS

COBALT

Cobalt is an essential constituent of vitamin B_{12}. Rumen microorganisms are able to synthesize all the B vitamins, but unless Co is present, they cannot synthesize B_{12}. Cobalt is therefore a required mineral.

Deficiency Symptoms - Vitamin B_{12} is required for energy metabolism, and so Co deficiencies begin as gradual poor performance. As the situation continues, the animal becomes very weak (lack of energy) and listless. Eventually, the animals become extremely emaciated. Actual cases of Co deficiency seem to be limited to Australia and New Zealand. In the U.S., the author would assume that subclinical cases occur, particularly in the Southeast (leached soils), where Co deficiency may appear as an aggravating factor to other trace mineral subclinical deficiencies.

COPPER

Copper is required for a vast number of enzymes, as well as for proper bone and hair formation, and the synthesis of hemoglobin. Not all the functions of Cu have been clearly identified as it has numerous interrelationships with other minerals. Copper is of considerable practical importance because there are areas in the western U.S. that are deficient. Also, sulfates and molybdenum readily combine with Cu, and can therefore reduce the availability of Cu. In many areas SO_4 or Mo concentrations are quite high, and induced Cu deficiencies are relatively common problems.

Deficiency Symptoms - The classical deficiency symptoms for Cu are lack of hair pigmentation and anemia. Anemia develops because of the necessity of Cu for the formation of hemoglobin.

The classical symptoms pertain primarily to simple deficiencies of Cu. Where high levels of SO_4 and Mo are creating induced deficiencies, the symptoms can be quite varied. Symptoms previously described have been severe diarrhea, muscular incoordination, pacing as opposed to walking, swayed backs and inability to use the hind legs, dilated pupils, and brain damage.

Toxicity Symptoms - Toxicity symptoms in the U.S. are usually the result of contamination as many insecticides contain high levels of Cu. Also, copper sulfate is used for a variety of purposes on farms and ranches; e.g. treatment of foot rot, etc., which can lead to contamination as well.

The clinical symptoms are similar to general liver failure. The liver will store excess Cu for a long period of time, and the animals will appear perfectly normal. When the liver reaches its threshold level (all it can hold), it will break down and very suddenly the animal will become acutely ill. As with most liver failures, jaundice will be apparent, and the animal will appear to be in intense abdominal pain. Death will come within a few days, depending upon the amount of stress placed upon the animal.

While diagnosis of Cu toxicity itself is somewhat difficult, the circumstances preceding the problem should not be. That is, if a pesticide or other contaminant is involved it should not be too difficult to find the general cause. Autopsy should reveal extremely high Cu levels in the liver and kidney.

There is no treatment, other than removing the animals from the source of contamination. As a general rule, whenever a number of animals suddenly become ill or die, a toxicity of some kind should be suspected, and the surviving animals moved to different feed and water.

IODINE
Iodine has long been known to be a required mineral, and likewise there are a number of areas (the Midwest in particular) that have been known to be deficient. The primary role of I is as a component in thyroxin, the main hormone produced by the thyroid gland.

Deficiency Symptoms -The classic deficiency symptom of I is an enlarged thyroid gland, commonly known as goiter. Reproductive problems have also been reported which include reduced male libido, irregular estrus, resorbed fetus, abortion, stillbirths, and hairless calves. The use of iodized salt has eliminated nearly all I deficiency problems in the U.S.

Toxicity Symptoms -Cattle can tolerate quite high levels of I, and so I toxicities do not occur under natural circumstances. However, organic iodide (EDDI) is used as a foot rot control, and overzealous or mistakenly excessive use always creates the possibility of a toxicity. Church (1974) reported an incidence of EDDI toxicity with sheep in which symptoms were non-specific; i.e. death, abortions, stillbirths, poor feeder performance, and failure of some of the ewes to breed the following season.

MANGANESE
    Manganese is used in a number of enzymes, and is known to be interrelated with several other minerals; i.e. other minerals can often substitute for Mn. Manganese is also known to be an essential element in bone and cartilage formation.

Deficiency Symptoms - In adult animals the only apparent symptoms are reduced estrus. Deficient calves will be born with swollen joints, knuckled over pasterns, and crooked legs. In feeder and stocker cattle poor performance would be likely, but it has been difficult to demonstrate.

Toxicity Symptoms - The toxic symptoms of Mn are simply poor performance. Cattle are able to tolerate quite high levels of Mn . . . toxicity problems seem to be more associated with other minerals that Mn ties up. For that reason extremely severe cases of overconsumption of Mn could result in a variety of symptoms.

MOLYBDENUM
    Molybdenum is believed to be an essential component of a number of enzymes. The actual requirement is not accurately known, but the quantity involved is so small that Mo is rarely, if ever, supplemented. The only practical problem Mo creates is when it is in excess, as it can tie up Cu and Zn and create induced deficiencies.

Deficiency Symptoms - The requirement for Mo is so low that clinical deficiency symptoms are not displayed even with purified diets.

Toxicity Symptoms - The primary toxicity symptom of Mo is the tying up of Cu. Clinical symptoms are quite varied and are discussed under the Cu section. In severe Mo toxicities, Zn has been reported to be tied up, and as discussed in the Zn section, typical Zn deficiency symptoms will be encountered.

ZINC
    Only a few functions for zinc have been specifically identified. It is clear, however, that Zn is required for a large number of

functions, since it is found in nearly all body organs and it is required at fairly high levels (30-40 ppm).

Deficiency Symptoms - Clinical symptoms of Zn deficiency do not ordinarily occur under natural conditions. Most feeds contain marginal amounts of Zn, and so subclinical problems (reduced growth) can be a problem. Also, excess Ca has been known to tie up Zn, which can also lead to subclinical deficiencies.

Experimentally produced deficiencies have resulted in dermatitis around the nose, lips, and later, the head and neck. In some cases (but not all) profuse salivation was reported. In the male animal, testicular degeneration and lack of libido have been reported.

SELENIUM

Selenium is a very important mineral as its deficiency and toxicity symptoms are quite severe, and both occur under natural

Figure 11-12. An apparent selenium deficiency. When mineral deficiencies/toxicities are encountered in stockyards or feedlots the history of the animals is often not accurately known. Accurate diagnosis is therefore usually difficult since many deficiency/toxicity symptoms are very similar, and can be produced either by simple deficiencies or excesses of individual minerals, or interactions between two or more minerals. However, with Se the symptoms are usually characteristic.

conditions. Making the matter more critical is the fact that the range between deficiency and toxicity is quite narrow; i.e. the requirement for most animals is about .01-.02 ppm, but levels as low as 5 ppm have been reported to create toxicities.

The actual function of Se has not been accurately determined. All that is known is that it has an interrelationship with Vit. E that serves to prevent nutritional muscular dystrophy (white muscle disease).

Deficiency Symptoms - Deficiencies of Se primarily affect young animals and the symptoms are quite characteristic. The condition is a muscular dystrophy, and hence the animals have a very difficult time walking or even standing. Typically the back will be hunched and the rear feet carried quite far forward and the front feet farther back than normal. Eventually the condition will lead to cardiac arrest. Autopsy will reveal lesions throughout the striated musculature to the point that the muscles look white (hence the name white muscle disease). Naturally occurring deficiencies are relatively common in areas where soil is of volcanic origin.

Toxicity Symptoms - Selenium toxicities have been reported since the 19th century, and continue to be a real problem in parts of the western U.S., Wyoming in particular. There appear to be two separate and distinct types of Se toxicities. Which toxicity the animal displays is apparently due to the form in which the Se occurs. Both conditions are severe and characteristic.

The first condition, known as alkali disease, occurs when the soil contains a high level of Se which results in the forage containing a high level of Se. Animals grazing the forage become anemic, and quite lame. The lameness is due to stiffness in the joints, and a severe overgrowth and deformation of the hooves. No other malady results in such severe hoof damage.

The second situation occurs where soil contains moderate levels of Se, but certain plants which accumulate Se are in the area. The grazing of these accumulator plants results in what is known as "blind staggers". As the name implies, the animal wanders around aimlessly. The animals go off feed, appear uncoordinated, and quite weak. In the final stages the animal becomes blind, and goes into paralysis. The disease can bypass the initial stages, and the cattle simply collapse, go into paralysis, and die.

IRON
Iron is an essential component of hemoglobin. While Fe is known to have other functions, the vast majority of the Fe in the

body is used for oxygen transport via hemoglobin. Most feeds are relatively high in Fe, and so under most circumstances it need not be supplemented, although iron oxide is routinely added to commercial mineral products. Iron oxide is a very poor source of Fe, but its addition to mineral supplements is only to give them the traditional reddish color, rather than an attempt to supply supplemental Fe.

Deficiency Symptoms - A lack of Fe decreases the amount of hemo-globin in the blood and anemia results. Clinical symptoms in adult animals are only a problem in very leached soils. Milk is quite low in Fe, and so there have been cases of calves receiving little feed other than their dam's milk developing Fe deficiency anemia. This situation, however, has primarily been restricted to dairy calves. Beef calves usually get plenty of iron either through grazing or eating dirt (dairy calves are often kept on concrete floors).

Toxicity Symptoms - Clinical toxicities with Fe do not occur under natural conditions. Subclinical problems have resulted in reduced performance due to high levels of Fe in water or pasture grass. Probably the most likely areas for such occurrences would be reclaimed strip mine areas, and pastures in the general vicinity of smelters, etc. Clinical cases have been produced experimentally, but the symptoms were unspecific.

OTHER TRACE MINERALS

There are a number of other trace minerals which are either not required, or are required in such small amounts that their necessity is not detectable without utilizing purified diets in exper-imental situations. The real importance of these minerals is that they can be quite toxic.

LEAD-Probably the most important mineral in terms of total num-bers of cases of toxicity and death is lead. Lead attacks the nervous system, and causes a number of neuromuscular problems. In the case of acute poisoning, the symptoms have been described as very similar to rabies. With chronic poisoning, cattle become very weak, stiff in the joints, and will eventually become inco-ordinated and go into convulsions.

The main cause of Pb poisoning is usually attributed to Pb paints; i.e. cattle chewing or licking on painted surfaces. Another problem has been areas near smelters, lead mines, etc. Quite recently problems have developed with automobile exhausts. Lead emissions from automobile exhausts have caused lead poisoning in cattle grazing vegetation near highways or urban areas with smog problems.

FLUORINE-Heavy intakes of F will cause a very severe softening of the teeth. It will also cause very abnormal bone growth (osteofluorosis) which can result in lameness and even immobility of the spine.

In times past the greatest cause of fluorosis was the utilization of rock phosphate as a source of supplemental phosphorous. Rock phosphate has a very high level of F, which must be removed if it is to be used as an animal feed. Once removed, defluorinated rock phosphate is a reasonably good source of P. Most stockmen and virtually all feed companies are aware of this, so this should no longer be a problem. Still, the stockman should be wary of bargain mineral products as P is the main ingredient in mixed mineral products, and "low fluorine rock phosphate" sells for ¼ to ⅓ the price of feed grade defluorinated rock phosphate.

More recently, F toxicity problems have occurred on pastures (from settled particulate matter) near aluminum and steel plants. While F is relatively high in the ground water of certain areas such as central and south Texas, very seldom is it high enough to cause toxicities. Effluent water from smelters, mines, etc., can contain enough F to create toxicities.

ARSENIC-Arsenic is well known as a poisonous substance, and occasionally cattle losses are attributable to As poisoning. There are no areas that the author is aware of, that have toxic levels of As in the soil or plants. Usually the source of As is in a pesticide. Cottonseed hulls and whole cottonseed have been implicated in Texas; i.e. pesticides used on cotton often contain arsenicals. Normally they cause no problem, but if pesticide instructions are ignored, contaminated cottonseed products can and have resulted. Toxicity symptoms are very similar to many other mineral toxicities; animals appear weak and incoordinated, and may go into convulsions.

MERCURY-Mercury is a mineral widely used in industry and agriculture which is quite toxic. Symptoms are similar to other mineral toxicities; weakness, trembling, convulsions, and sometimes excessive salivation. The greatest danger in the poisoning of cattle comes on farms where access to seed grain might accidentally be obtained, as seed grain is usually coated with Hg or other toxic compounds to avoid mold and other fungal losses.

MINERAL SUPPLEMENTATION

In most situations beef cattle require some type of mineral supplementation (other than salt). This is essentially due to two reasons:

1. Different parts of plants contain different concentrations of Ca and P. Grains contain relatively high levels of P, but are practically devoid of Ca. Grasses, plant stems and leaves usually contain adequate levels of Ca, but are normally low in phosphorous. Therefore cattle on pastures normally have to be supplemented with P, and cattle on grain rations need supplementary calcium.

2. As a general rule, plants have relatively low requirements for trace minerals. The level of trace minerals found in plants will vary greatly with the content of the soil. In areas grazed over a long period of time, the trace mineral content of the soil will become depleted due to continuous removal.

MINERAL SUPPLEMENTATION ON IMPROVED PASTURE

Since grasses are typically low in P, P must usually be supplemented. Humid areas such as the Southeast, which have acid soils, are also sometimes low in Ca, and so Ca is sometimes called for. If the pasture is a grass-legume mix, Ca may not be necessary since legumes are usually quite high in Ca. As a practical matter, nearly all sources of P also contain Ca, and so the two usually are supplemented together.

When pastures are fertilized, they normally receive only N, P, and K. Typically they are not fertilized with trace minerals since grasses do not usually respond to trace mineral fertilization. For this reason it is usually best to supplement cattle on improved pastures with trace minerals.

Pastures fertilized with manure may receive some trace minerals if the animals generating the manure were supplemented with trace minerals. However, this is not a very reliable source. In humid areas, even if adequate trace minerals are contained in the manure, they can be leached out fairly quickly.

MINERAL SUPPLEMENTATION OF RANGELAND

Range supplementation is essentially the same situation as improved pasture in that P is almost always going to be low. Ca is often entirely adequate, particularly in the southwestern United States.

The area of question is with trace mineral supplementation. For the most part, the rangelands in the more arid regions of the western U.S. have mineral soils with relatively high levels of trace minerals. Plant growth in these areas is relatively slow (due to the arid conditions), and so the trace minerals in the soil for the most part have not been depleted. Many ranchers have apparently gotten by adequately with only salt and bone meal (Ca & P). The key to that statement . . . is the word "apparently". As mentioned earlier, the deficiencies that hurt the cattle industry the most are the ones that can't be seen. For this reason, it is probably a good practice to provide trace minerals to range cattle as the cost is very little (in addition to P), and there is always the possibility that one or more trace elements will be in inadequate supply.

MINERAL SUPPLEMENTATION IN THE FEEDLOT

Mineral supplementation in the feedlot will obviously depend upon the type of ration fed. Feedlot rations are typically high grain rations, but the amount and type of roughage and other ingredients fed can change supplementary mineral requirements considerably. Then too, high roughage rations are sometimes fed, which can also radically change the type of mineral supplementation required; e.g. corn silage is notoriously low in all minerals and therefore high silage growing rations would require both Ca and P as well as trace mineral supplementation.

Grain, as mentioned earlier, is relatively high in P, but practically devoid of Ca. For this reason supplementary Ca, usually in the form of limestone, is added to feedlot rations. Phosphorous is also usually added to feedlot rations, but in not nearly as great of quantities as Ca. Normally 6 to 12 times as much Ca will be added.

Sulphur is almost always added to feedlot supplements since urea or other forms of NPN are typically relied upon for a substantial amount of the ration nitrogen (protein). Sulphur is contained in 3 amino acids (methionine, cystine, and cysteine), and if it is not supplemented, the rumen microorganisms could not synthesize those amino acids (from NPN).

While most texts state that potassium need not be supplemented, it is routinely added to feedlot rations by most consulting nutritionists, particularly with milo rations. Responses to K are no doubt due to the lower levels of roughage that are now used in most feedlot finishing rations (green roughages are typically high in K).

Sodium (in the form of salt) is usually either added in the ration or made available on a free choice basis. Many feedlots feed salt as a palatability agent, or to increase water consumption for prevention of urinary calculi (see page 111).

Trace mineral supplementation depends somewhat upon the roughage and other ingredients mixed in with the grain. If alfalfa is used as the roughage source, trace mineral supplementation is not nearly as critical as with most other commonly used forms of roughage. Molasses contains considerable amounts of trace minerals, and like alfalfa, can reduce the necessity of trace mineral supplementation. Milo generally has slightly more trace minerals than other grains, and in many southwestern feedlots, rations of milo, alfalfa hay, and molasses are routinely fed without any trace mineral supplementation whatsoever. Hale (1978)* of the University of Arizona has researched the use of trace mineral supplementation on milo, alfalfa hay, and molasses rations, and found responses only to Co. The author would like to comment, however, that this information would probably only apply to native yearling cattle. Cattle coming from humid regions, such as the Southeast, would probably respond to nominal trace mineral supplementation.

VITAMINS

Rumen microorganisms have the ability to synthesize all the B vitamins, which eliminates their dietary necessity and greatly reduces the complexity of supplementing beef cattle with vitamins. There are only three vitamins that are commonly supplemented; vitamins A, D, and E. Only with vitamin A are clinical deficiencies relatively common. It is possible to induce clinical deficiencies with vitamin E, but the feeds required to do that would result in a general malnutrition (unless done experimentally with purified diets). These and other vitamins of practical importance will be discussed individually.

VITAMIN A

Vitamin A is the most important vitamin in beef cattle nutrition. It is a fat-soluble vitamin and it is not found in feedstuffs of plant origin. There are, however, a group of compounds (carotenes) which are found in plants and can be converted to vitamin A in the animal's body. Green forages usually contain relatively high levels of carotene, whereas grains and most other concentrates fed to cattle are extremely low in carotene.

This makes cattle dependent upon green plants for vitamin A (carotene). As plants mature or otherwise turn yellow or brown,

*Personal communication.

the amount of carotene present decreases rapidly. Dry, mature plants are essentially devoid of carotene.

Animals are able to store vitamin A in large amounts in the liver, and mature animals in good condition can go 3-4 months on deficient diets without supplemental vitamin A. This mechanism allows cattle to get through the worst of winter and summer periods when the grass is dormant. If bad weather keeps the grass dormant longer than usual, the cattle may exhibit clinical signs of deficiency if not supplemented. Stress can rapidly use up vitamin A stores, and thereby effectively reduce the time required for deficiencies to occur. It is therefore a good idea to supplement vitamin A anytime cattle do not have access to green grass.

Carotene is relatively unstable and is readily denatured by heat and light. Hay and silages are therefore often deficient in vitamin A activity utilizable by the animal. For this reason, feedlot animals are usually supplemented with vitamin A since silages and hays are normally their only source of forage. Usually this is accomplished via a pelleted protein, vitamin, and mineral supplement mixed in with the ration. Also, some feedlots routinely inject their cattle with an injectable preparation of vit. A.

Deficiency Symptoms - The classic deficiency symptom of vitamin A is usually reported to be night blindness. However, an observant stockman should be able to notice lackluster haircoat long before any night blindness occurs. But as mentioned, night blindness is considered to be the first clinical symptom of vitamin A deficiency. Cattle will stumble and bump into things in dim light. The condition is unique to vitamin A. In the past, this condition was known to "old timers" as cottonseed meal poisoning, as the first feedlots in the Southwest fed rations consisting of nothing more than cottonseed meal and hulls. The people in the business back then knew that cattle could not be held on those rations more than 120 to 150 days or they would develop "meal staggers". What was believed to be a toxic effect of the cottonseed meal is, of course, now known to have been a vitamin A deficiency.

As the deficiency continues, young animals may become completely blind. Vitamin A is also used to maintain mucous membranes, and so abnormal appearances will result. The eyes may show excess tearing, and the nose and throat may have a mucal discharge. Reportedly as a secondary problem cattle will be more susceptible to respiratory infections.

In further progression of vitamin A deficiency, young cattle will have profuse diarrhea, and will intermittently go into convulsions. Feedlot animals will develop what is known as anasarca, which is a swelling of the lower legs and brisket. Cows will abort or fail to conceive, and bulls will have reduced fertility. Eventually, cattle can die of vitamin A deficiency.

Toxicity Symptoms - Intakes in excess of 50 times the requirement over a period of time can eventually create a toxicity. The symptoms are many and varied, but include copious mucous discharge, loss of hair, loss of appetite, swelling of the sinuses, and spontaneous fracture of the bones.

While toxicities have been produced experimentally, as a practical matter they are not a problem. If a toxicity were to occur, it would have to be the result of a miscalculation or mixing error with synthetic vitamin A. A toxicity from excess carotene in natural feedstuffs is unlikely. Conceivably a toxicity could be created with a gross excess of injectable vitamin A.

VITAMIN D

Originally known as the "antiricketic factor", vitamin D is required for the absorption and mobilization of Ca. Without vitamin D, rickets will result even if adequate Ca and P are in the diet. The primary dietary source of vitamin D activity for cattle is ergosterol, a vitamin D precursor contained in plants. The compound is activated when it comes in contact with sunlight (ultraviolet light). Hence, most field-dried hays contain active vitamin D.

If the animal has access to sunlight, it really doesn't need a dietary source of vitamin D, since sterols produced in the skin can be activated (by sunlight) to form vitamin D. Since beef cattle typically are kept outdoors, vitamin D deficiencies are very rare.

Deficiency Symptoms - The only deficiency symptom that has ever been demonstrated for vitamin D has been rickets. It is conceivable, however, that osteomalacia could be produced in adult cattle kept indoors. The time required would no doubt be quite extensive.

Toxicity Symptoms - Vitamin D can be toxic in amounts exceeding 10 times the requirements. There are some non-specific symptoms such as scouring, depressed performance, and lethargy. The most serious symptom is calcification of soft tissues. As with other vitamins, natural toxicities are highly unlikely. Mistakes in the utilization of synthetic substitutes are the only real avenue for clinical toxicities.

VITAMIN E

The actual physiological functions of vitamin E are not clearly understood, but it is known to function with Se in the prevention of nutritional muscular dystrophy (white muscle disease). Also, growth responses in feedlot cattle have been reported.

Deficiency Symptoms - Reproductive problems have been reported in other species of animals, but have not been clearly documented in beef cattle. The classic deficiency is a muscular dystrophy in calves. The skeletal and cardiac muscles take on a whitish hue, and hence the term, "white muscle disease".

Toxicity Symptoms - Animals can evidently tolerate a very high level of vitamin E, and so toxicity is not of practical importance.

VITAMIN K

Vitamin K can be synthesized in the rumen and gut, so it is generally not of practical importance. There is one exception to this: certain molds that grow on sweet clover produce an antagonist of vitamin K called dicoumeral. Ingestion of molds grown on sweet clover can thereby create a vitamin K deficiency. Vitamin K is required for normal blood clotting; thus a deficiency can result in severe hemorrhaging.

B VITAMINS

The B vitamins comprise a large group of water-soluble vitamins that are essential for animal life. Even so, they are of limited practical importance in beef cattle, since the rumen bacteria can synthesize all of the B vitamins.

There are instances, in which feedlot and sometimes pasture cattle develop neurological problems that will respond to thiamin (vit. B1) supplementation. The condition is called polioencephalomalacia. Injection with thiamin will completely reverse the condition, if done in time. Untreated animals will usually die. Symptoms include blindness, convulsions, necrosis of the brain, and a body position in which the head and neck are bent around backwards. The actual cause of the problem is not known, but it is thought not to be a simple thiamin deficiency; rather, the result of some kind of thiamin antagonist.

The only other time B vitamins become of practical importance is in cattle that do not have a functioning rumen; i.e. (orphaned) calves and sick cattle. For this reason good milk replacers contain ample B vitamin supplements, and cattle that have been sick for a period of time will often respond to B vitamin injections.

CHAPTER 12   UREA AS A FEEDSTUFF

To most cattlemen, urea is considered nothing more than a necessary evil; i.e. because of its high protein equivalency, it can "cheapen up" a protein supplement or ration, but actually it is not as good as all natural protein. This is not necessarily true.

It is true that urea is often overutilized, but when used moderately it actually enhances most protein supplements. This is because urea is broken down very quickly, providing a readily available source of nitrogen for the rumen microorganisms. Many natural protein supplements are broken down relatively slowly, and therefore do not provide optimum levels of nitrogen for microbial use. Judicious use of moderate amounts of urea often enhance the value of natural protein supplements.

The cattleman's distaste for urea usually stems from instances of urea toxicity. Urea, like most synthetic substances, must be used in moderation. When used in excess, it can produce toxicity and death. But toxicity is a gross manifestation of the overuse of urea. A far more common problem associated with the overuse of urea is simple reduction in growth or performance (as opposed to a more moderate level of urea).

UREA RECYCLING

Understanding urea recycling is basic to an understanding of ruminant nutrition. Urea is not just a man-made synthetic source of protein (nitrogen), but it is also a metabolic product formed in the body.

When rumen microorganisms degrade proteins, the nitrogen released is quickly hydrolyzed (hydrogen is added) to form ammonia (NH_3). If more nitrogen (NH_3) is present than the microorganisms can utilize, the excess will be absorbed into the bloodstream (through the rumen wall) as ammonia (see Figure 12-1, point 1).

Once in the bloodstream, the ammonia is taken to the liver (all blood leaving the rumen goes directly to the liver) (Figure 12-1, point 2). One function of the liver is to detoxify the blood, and since ammonia is toxic to the animal, the liver synthesizes it into a benign compound. The compound formed is urea, which is then released in the bloodstream.

The urea is transported to the salivary glands (Figure 12-1, point 3), where it is absorbed and recycled into the rumen via the salivary secretions. Some urea is also absorbed through the

222

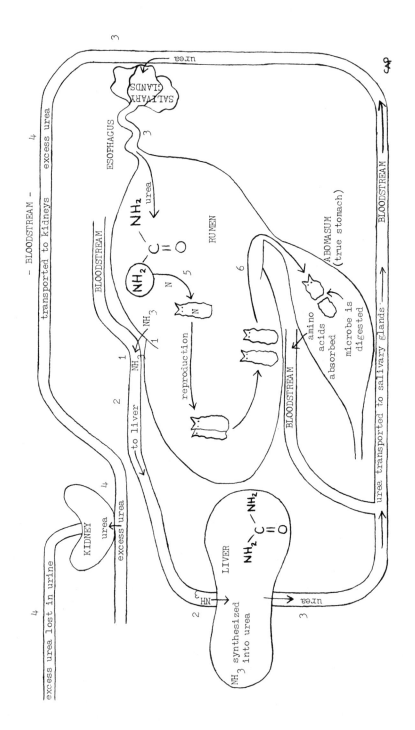

Figure 12-1.   UREA RECYCLING

1. Ammonia (NH_3) is released by the microorganisms and is absorbed through the rumen wall.

2. Ammonia absorbed through the rumen wall is transported to the liver where it is synthesized into urea.

3. Urea is then transported from the liver to the salivary glands, where it is recycled into the rumen via the salivary secretions.

4. Excess urea is transported to the kidney, where it is lost in the urine.

5. Microorganisms break away the amine groups from the urea, and use the nitrogen to form their own proteins. These proteins are then used for the microorganism populations to multiply.

6. Microbe is transported to the true stomach and small intestine and is digested as protein. The amino acids released are then used by the ruminant to form its own proteins.

rumen wall from the bloodstream. . Excess urea is transported to the kidneys, where it is excreted in the urine (Figure 12-1, point 4).

The urea that is recycled into the rumen (via the saliva and rumen absorption), is used as a nitrogen source (protein source). The microorganisms then break away the amine group (NH_2), and form their own proteins (Figure 12-1, point 5). As in natural protein utilization, the microbes use this synthesized protein to facilitate their own growth and reproduction. This increases the total microbial population in the rumen. Some of the microbial cells are washed out of the rumen and ultimately are transported to the abomasum (true stomach) and small intestine. There they are digested by proteolytic enzymes (Figure 12-1, point 6), and the proteins contained in the microbes are then broken down into their constituent amino acids. These amino acids are then absorbed and used by the ruminant to form its own proteins.

UREA TOXICITY

The term "urea toxicity" is actually a misnomer as urea itself is not toxic. From the previous discussion, this should be clear, as urea is formed in the liver and released into the bloodstream. Urea toxicity is actually a condition caused by the end product of urea degradation; i.e. NH_3, when too much urea is introduced into the rumen. Properly named, the condition caused by excessive amounts of urea and some other non-protein nitrogen compounds is actually ammonia toxicity.

As pointed out in the preceding section, excess N in the rumen is given off as NH_3, absorbed into the bloodstream, and transported to the liver. The liver then synthesizes the NH_3 into urea and releases it back into the bloodstream.

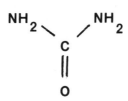

Figure 12-2. UREA MOLECULE.

A toxicity problem occurs when more NH_3 is released into the bloodstream (from the rumen) than the liver can synthesize into urea. With natural proteins this rarely becomes a problem because natural proteins are usually fairly difficult for the rumen micro-organisms to break down. Ammonia or nitrogen released is therefore spread over a period of time.

Urea and several other non-protein nitrogen compounds are broken down much more quickly. Therefore when excessive amounts of urea are introduced into the rumen, a large percentage of the N contained in the urea is given off as NH_3 in a short period of time. The amount of NH_3 given off can be more than the liver can synthesize into urea. When this situation occurs, the NH_3 is left in the bloodstream. Toxicity symptoms will occur when the ammonia reaches the brain and other nervous tissues.

Symptoms will occur shortly after the overdose, and will include tremors, muscle incoordination, staggers, convulsions, and death, in that order. There are a number of other maladies that will create somewhat similar symptoms: grass tetany, organophosphate poisoning (adverse reaction to grubacides), anaphylactic shock (adverse reaction to injections), and to a certain extent, nitrate toxicity. The circumstances prededing the toxicity should indicate the problem involved. If in doubt, however, smell the animal's breath. The physiological defense mechanism is to exchange NH_3 at the lungs (to remove NH_3 from the bloodstream). An animal affected by urea toxicity will therefore have a breath that smells of ammonia.

If a lethal dose of urea is ingested, death will usually occur from 20 minutes to 2 hr. If the animal lives for more than 2 hr, it will usually survive. Field treatment consists of drenching with at least one gallon of vinegar. If administered prior to the convulsive stages, survival rate is normally very good, and apparently without aftereffect.

According to Church et al. (1974) .18-.227 grams/lb bodyweight are required to kill cattle in "poorly fed or starved" condition, and about .29-.34 grams/lb for well fed cattle, when consumed in 30 minutes or less. In other words, if a protein supplement contained a 30% crude protein equivalency from urea, it would take from 3.3 to 4.2 lb of the supplement to kill a poorly fed 900 lb cow, and from 5.3 to 6.3 lb to kill a well fed 900 lb cow.

Feedlots normally use supplements high in urea, and cases of toxicity typically occur when excessive amounts of supplement are accidentally fed. In feedlots that mix their own supplements, human and mechanical malfunction of the metering equipment have led to urea toxicity. In the range situation, toxicities have been caused by aggressive animals eating more than their share of hand fed supplements, as well as simple overfeeding of high urea supplements. With self limiting supplements, such as liquid feeds, toxicities have occurred when the wrong formulation is used; i.e. most range liquid supplements use phosphoric acid as a control agent, and when insufficient levels are included, greater than the desired intake can result.

REASONS FOR OVERUSE

On a weight basis, urea normally costs about 10 to 40% more than cottonseed or soybean meal. But since it has about 6 times the crude protein equivalency, urea is used extensively to reduce the costs of protein supplements. Virtually all beef cattle protein supplements and mixed rations contain urea.

The addition of a small amount of urea, because of the 281% crude protein equivalency, can greatly increase the crude protein equivalency of a feed. For instance, if a protein supplement has a 40% crude protein rating, the addition of 14.5% urea will double the crude protein of that supplement (80.7%). If a complete feed contains 10% crude protein, the addition of only 1.5% urea will increase the crude protein of that feed to 14.2%.

The crude protein equivalency of ruminant feeds is usually represented as "the protein content", of that feed. Most state feed laws also require that the equivalency coming from non-protein nitrogen be specified. But for the most part, ruminant feeds are normally represented to farmers and ranchers as 30%, 40%, 80%, etc., protein products. Knowing that one feed product may be bought over another one because it is higher in "protein" (has a higher crude protein equivalency), many feed manufacturers are tempted to add excess urea.

The use of urea can also allow the feed manufacturer to substitute lower quality feedstuffs for higher quality natural proteins. Indeed, some products sold as protein supplements contain essentially no natural proteins (liquid supplements are particularly notorious for this). For example, a good quality range supplement (for most conditions) would have approximately a 40% crude protein rating, with less than 1/3 of that protein rating in the form of non-protein nitrogen. By weight then, that supplement would contain about 4.5% urea and somewhere between 60 and 70% oil meal (cottonseed meal, soybean meal, etc.). By doubling the amount of urea; by adding 9% urea instead of 4.5%, the amount of oil meal could be reduced to almost half what was contained in the initial product and still maintain the 40% crude protein rating. This allows the feed manufacturer to add 30 to 35% "filler" to the product; i.e. some type of low quality, inexpensive grain by-product or roughage.

There are many interacting conditions that dictate how much urea can be utilized in a ruminant feed. When those conditions are quantified and the "correct" amount of urea is used, results will be "good".

FACTORS AFFECTING UREA UTILIZATION

The most well known factor affecting urea utilization is the necessity of having a source of readily available carbohydrate. The carbohydrate forms the base to which the rumen microorganisms add the NH_2 (taken from the urea molecule), to form amino acids (Figure 12-3). If there is not sufficient readily available carbohydrate present, then amino acids cannot be formed, and the NH_2 will be lost as NH_3 through the rumen wall.

Figure 12-3. RUMEN AMINO ACID SYNTHESIS.

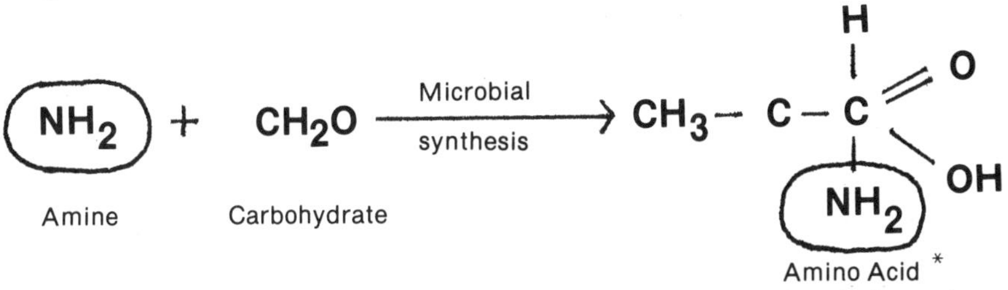

$$\boxed{NH_2} + CH_2O \xrightarrow[\text{synthesis}]{\text{Microbial}} CH_3 - C - C \overset{\displaystyle O}{\underset{\displaystyle OH}{\diagup}}$$

Amine   Carbohydrate   Amino Acid *

*This particular amino acid is alanine, which has the simplest chemical structure of all the amino acids. All amino acids contain at least one amine group (NH_2), from which they get their name.

A situation where there is little readily available carbohydrate would be on a high roughage ration, such as hay or range grass. While there is energy contained in these roughages, it cannot be released by the rumen microorganisms very fast. On high grain rations there is much more readily available carbohydrate, and therefore higher levels of urea may be used to substitute for natural proteins.

The amount of soluble nitrogen contained in the natural protein portion of the ration also plays a major role in the amount of urea that may be used as a protein supplement. Some feeds, high-moisture grains in particular, contain a large percentage of their total nitrogen in the form of non-protein nitrogen. For instance, Prigge et al (1978) found that 4.3% of the nitrogen contained in dry-rolled corn was in the form of soluble non-protein nitrogen, whereas ground high-moisture corn contained 56.6% of its nitrogen in a soluble non-protein nitrogen form. Obviously only a limited amount of urea can be added to a feed that already has high concentrations of soluble non-protein nitrogen.

Adaptation of the animal will limit the amount of urea that can be used in the beginning. It takes approximately 10 days to 2 weeks for the animal to become adapted to fully utilize urea. Feeding a large amount of urea to unadapted cattle will result in a substantial loss of the nitrogen via the urine. Likewise, the level of urea required to create toxicity is reduced.

While it takes about two weeks for cattle to develop an adaptation to urea, the adaptation can be lost in a much shorter period. Gallup (1953) found that steers could lose their adaptation in just 48 hours. For this reason, once cattle are started on high levels of urea, they should be given a constant daily intake of urea.

Because of the adaptation time required, feedlot cattle should be brought up on urea gradually. As it is practiced in the feedlot industry today, most cattle are given a "starter supplement" which is typically lower in non-protein nitrogen and higher in natural proteins than supplemental premixes used during the bulk of the feeding period. Theoretically, adaptation time would be a factor in feeding range cattle, but the timing of intake is a factor that greatly overshadows the importance of adaptation time.

To fully understand the effect of the timing of intake, one needs to remember how rapidly urea is broken down (and NH_3 released). Within 50 minutes after ingestion, almost all urea will have been broken down by the rumen microflora (Davis, 1953). This means that the amount of urea that can be fed will depend upon how often the cattle have access to the urea-containing feed.

Under normal circumstances the amount of NH_3 released in the rumen that an animal can utilize remains fairly constant. Graphically it is represented in Figure 12-4. The vertical axis represents the amount of NH_3 released in the rumen, and the horizontal axis represents time. The line drawn across the graph symbolizes the maximum level of NH_3 that can be utilized at any given time. Ammonia released in excess of that level would be absorbed through the rumen wall and eventually lost with the urine.

In the feedlot situation, most cattle will receive about 1 lb of a supplement that will run approximately 40-60% crude protein. Typically urea will make up from ½ to ⅔ of the protein equivalency in those supplements.

Feedlot cattle will normally come up to the feedbunk and eat from 5 to 10 times during the day. If the level of urea contained

Figure 12-4. AMMONIA RELEASE IN RUMEN FROM UREA.

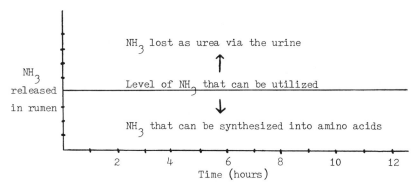

in the ration is in line with other factors affecting NH_3 utilization, a theoretical graph of the ammonia release would look like Figure 12-5. In this case there are six separate NH_3 release curves which would represent six feedings at the bunk. Spread over a period of time, the NH_3 released never becomes excessive and the animal is able to utilize efficiently the urea it ingests.

Figure 12-5. FEEDLOT AMMONIA RELEASE IN RUMEN FROM UREA.

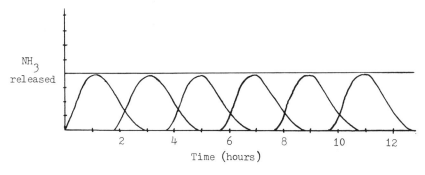

If an attempt is made to feed the same amount of urea as a supplement to grazing cattle, an entirely different situation prevails. Instead of being able to consume the supplement several times a day, range or pasture cattle normally have access to supplements a maximum of only once or twice a day. If the supplement is self fed, the cattle will normally consume it in the early morning and late afternoon when they come to water. The NH_3 release patterns would look like Figure 12-6. Rather than having multiple feeding periods, the animals consume the urea only twice during the day.

Therefore, even though the same level of total urea was fed, and the total amount of NH$_3$ released was the same, the timing of intake concentrated the NH$_3$ so as to create a loss of NH$_3$ (and a stress upon the animal). Aggravating the condition, of course, is the fact that range grass contains much less readily available carbohydrate than feedlot rations, and the ability of the rumen microbes to utilize the NH$_3$ is reduced further. This means that the horizontal line representing how much urea can be utilized would be lowered on Figure 12-6. But as mentioned earlier, the carbohydrate problem is greatly overshadowed by the timing of intake.

Figure 12-6. RANGE AMMONIA RELEASE FROM UREA IN RUMEN

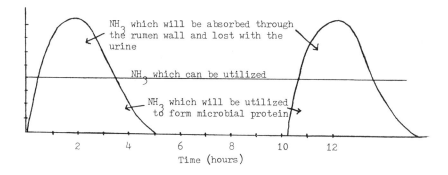

Considerable time and detail have been given to this problem because it is a very common situation. It is commonplace to see pasture and range cattle receiving supplements containing levels of urea (or other non-protein nitrogen compounds) that would be suitable only for feedlot use. Stated simply, the problem is this:

1. Feedlot cattle typically receive and consume 0.5 to 1.5 lb of a 30-80% CP high urea supplement over a 24 hr period.
2. Range cattle often receive and consume 0.5 to 1.5 lb of a 30-50% CP high urea supplement in about a 30 minute period.

As stated earlier, the reader should not interpret the discussion of the problems associated with excess urea supplementation to be a condemnation of the use of urea. Likewise, it should not be interpreted that urea shouldn't be used for range supplementation. When used in proper amounts, urea can be used quite well as a partial substitute for natural proteins in a range supplement.

A widely used formulation that has given good results on pasture and range cattle consists of 4% urea, combined with 75 to 80% oil meals, with the remaining fraction consisting of salt and trace minerals.

Natural proteins are broken down to form NH_3, just as urea is, but the time required is much greater. Soybean meal has been shown to release NH_3 at a fairly uniform rate for 12 hr (Davis and Stallcup, 1967). A supplement such as the one previously described would therefore have a release pattern similar to Figure 12-7. Theoretically, NH_3 released from the urea takes care of the immediate requirement for nitrogen, while NH_3 from the oil meal is released much more slowly, thereby taking care of the longer term requirement for nitrogen. In reality, of course, there would be only one NH_3 release curve, which would probably look similar to Figure 12-8. A small amount of excess NH_3 is represented here, which would occur occasionally, depending upon consumption of the supplement, cattle adaptation, etc., etc. However, the excess is not nearly as great as the high urea supplement represented in Figure 12-6. Also, NH_3 is available for microbial use continually with the moderate level urea supplement, whereas

Figure 12-7. AMMONIA RELEASE OF COMBINATION SUPPLEMENT UNDER RANGE CONDITIONS.

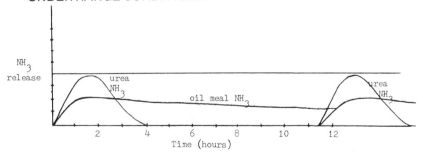

Figure 12-8. COMBINED AMMONIA RELEASE OF NATURAL PROTEIN AND UREA UNDER RANGE CONDITIONS.

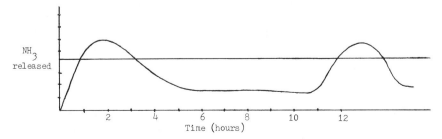

with an all or high urea supplement, there would be an NH_3 deficit between feedings.

The purpose of the section has been to point out the different variables that will affect the level of urea that may be used. Special attention has been given to the effect of timing of intake. The graphical illustrations have been greatly simplified and represent concepts rather than actual physiological data. The simplification has been for the sake of clarity, since this is a problem that every stockman should be aware of as it causes serious production losses in pasture and range cattle.

PRODUCTION LOSSES CAUSED BY THE MISUSE OF UREA

The toxic effects of urea are the major concern of most cattlemen. However, non-clinical problems cause far more losses to the cattle industry than what few cattle are actually killed. The losses come in the form of lost production. In feedlot and stocker cattle the losses consist of reduced weight gain.

The most serious problem associated with the overuse of urea is that the animal simply does not receive enough protein. The outward effect, of course, is reduced production. Sorting out how much reduction is due to protein deficiency, and how much is due to stress and energy loss is difficult. The author is not aware of any research work clearly delineating these different effects. There are, however, exhaustive amounts of work showing reduced weight gain due to protein deficiencies. If nitrogen (from urea) that has been calculated to supply part of the animal's crude protein requirement is lost in the urine, then clearly overdependence on urea can cause reduced weight gain due to protein deficiency.

A more difficult production loss to quantitate is the reduction in reproductive efficiency in brood cows. As pointed out on page 277, protein has been shown to have no direct effect upon the estrous cycle. Energy is considered to be the limiting factor in initiating estrus. But as also pointed out on page 29, protein is required in order to make energy available from low quality roughages. Therefore, when protein intake is inadequate (or reliance on urea excessive), cow herds grazing low quality roughages will have lowered calf crop percentages.

This problem has become more serious in recent years. Nutritionists have long advocated the feeding of supplementary protein to cow herds grazing low quality roughages. Cattlemen have bought commercial products containing heavy concentrations of non-protein nitrogen and have found them to be of little value.

Many have therefore concluded that protein is not effective and that grain (energy) should be supplemented instead.

As explained on page 31, grain makes a poor range supplement. The reason for the misunderstanding is that just because a product is labeled Protein Supplement and has a high crude protein rating, that does not mean that it will be effective. Fed correctly, protein supplements are very effective at increasing reproductive efficiency of cows grazing poor quality forages.

CONCLUSIONS

To some it may seem that the discussion of urea has covered too much detail; that a discussion of this sort would be better suited to a nutrition text. However, in field work it has been the author's experience that non-clinical problems with urea are a major cause of production losses. The greatest losses appear to be with range cattle.

To summarize the material presented it can be said that urea is a nitrogenous compound that is broken down in the rumen very quickly. If excessive amounts are fed, toxicity and death can occur. If fed in excessive but sublethal amounts, ruminant animals will undergo stress, and can suffer protein deficiencies (when excessive amounts of urea are fed in the hopes of meeting the animal's crude protein requirement).

Nothing in this discussion should be construed to imply that urea is an undesirable feed ingredient. Urea, when used properly, is a good feed ingredient. The purpose of this section has been to outline the details or factors that will influence the level of urea that can be used. These factors include the amount of readily available carbohydrate, the amount of soluble nitrogen and naturally occurring non-protein nitrogen compounds in the basal ration, the timing of intake, and the adaptation of the animal.

CHAPTER 13   SOME PRINCIPLES OF GENETICS

In order to more completely understand the breeding of livestock, one must understand a few basic principles of genetics.
Only those principles which have direct application to beef cattle
breeding will be presented here.

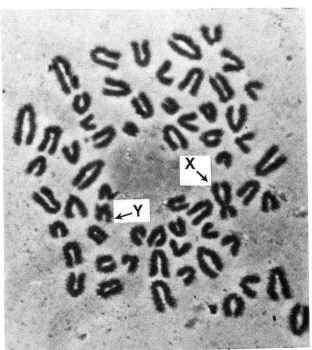

Figure 13-1. Cattle chromosomes (male) in a white blood cell, magnified
approximately 1,000 X. Genes, the chemical entities that actually
control heredity, are carried on the chromosomes. (From Keiffer
and Cartright, 1967).

GENES
    Genetics is defined as the study or science of inheritance.
Genes are the actual chemical entities that transmit the inheritance
traits. Genes are carried on the chromosomes, which are the
strands of genetic material that are carried with each individual
cell at the time of its division. Chromosomes come in pairs, so
that as cells divide, half of the genetic material goes with the
new cell, ensuring that it will be a perfect duplication of the parent
cell (in asexual reproduction). (At times, however, chromosomes
can become twisted or misplaced, which leads to mutations.)

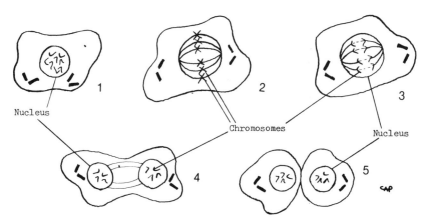

1. Cell before division.
2. Division begins, chromosomes are paired up.
3. Chromosome pairs are separated.
4. Separate cell nuclei form.
5. Separate cells emerge.

Figure 13-2. ASEXUAL CELL DIVISION.

In sexual reproduction the reverse of this situation occurs. Instead of two chromosomes from the same source dividing, two different chromosomes from different sources come together. That is, one chromosome comes from the male, and one from the female . . . and together they form a pair. Thus a cell is formed that is similar yet different than either parent.

More specifically, the cells of each animal will contain a given number of chromosome pairs; cattle have 60 pairs. When the animal forms gametes (ova and sperm cells), one chromosome from each pair is carried by the gamete or germ cell. Thus, sperm and ova cells each contain 60 single chromosomes. When fertilization occurs and the sperm enters the ova, the 60 single chromosomes from each germ cell will seek out its matching chromosome and form 60 chromosome pairs.

DOMINANT AND RECESSIVE GENES
Some genes display what is known as dominant or recessive action. When the chromosomes from sexual reproduction come together, there will be two genes (one from each chromosome) which control the same trait. If one of the two genes is a dominant gene, the trait it represents will be the one the offspring will express. The trait represented by the recessive gene will not be expressed, as long as a dominant gene is present.

The types of traits controlled by dominant and recessive genes are usually what are termed qualitative traits. These are traits that can readily be seen; e.g. coat color, horns or polled, etc. For instance, the white face of Hereford cattle, is controlled by a dominant gene. Thus, when Hereford cattle are mated with other breeds, the offspring will always have a white face. This is true not only because the white face is dominant, but also because Hereford cattle are evidently "homozygous" for the white face gene; i.e. they carry no other gene for face color. There are normally only two genes that control any one qualitative trait. Thus, there are usually only two alternatives within a breed. Cattle usually have only two color patterns per breed, are horned or are polled, etc.

Symbolically, a capital "D" could represent the dominant gene, and a small "r" could represent the recessive gene. Assuming an equal distribution of genes, four combinations of genes could be present in any one animal; i.e.

combination #	genotype
1	DD
2	Dr
3	rD
4	rr

Combinations #1 and #4 (DD and rr) are said to be homozygous because the animal carries only one gene for that trait. Combinations #2 and #3 (Dr and rD) are said to be heterozygous because they contain both the dominant and recessive gene. For instance, the black coat color in Angus cattle is a dominant gene. If an even distribution of genes is assumed (which is not true), it could be expected that 75% of all Angus cattle would be black and 25% would be red.

Whenever the appearance or physical attributes of an animal are referred to, they are described as the animal's phenotype. Phenotype being something that can be seen or measured. Whenever the genetic makeup of an animal is referred to, it is known as the animal's genotype. The genotype is something that cannot always be determined. For instance, in the case of combinations #1, #2, and #3, it cannot be determined from physical appearance if the animal is homozygous or heterozygous, since black coat color is dominant. But in combination #4, it is known that the animal is recessive homozygous since the color red cannot be expressed if a dominant gene (D) is present.

To see if an animal is homo or heterozygous for a trait, it can be bred with animals that are known to be recessive homozygous. If the animal in question is truly homozygous, then all the offspring will display the dominant trait; i.e. all the offspring will be heterozygous.

$$D_1D_2 + r_1r_2 = D_1r_1$$
$$D_1r_2$$
$$D_2r_1$$
$$D_2r_2$$

If the animal in question is heterozygous, 50% of the offspring will display the recessive trait; i.e. be recessive homozygous:

$$Dr_1 + r_2r_3 = Dr_2$$
$$Dr_3$$
$$r_1r_2$$
$$r_1r_3$$

Animals that are homozygous and are able to always pass their phenotypical traits on to their offspring are said to be prepotent or true breeders.

GENE FREQUENCY

Gene dominance plays a very important role in determining what the qualitative traits of animals will be, but there is another factor that can be of even greater significance, the concept of gene frequency. As the term implies, gene frequency deals with the frequency by which genes are encountered within a population of animals.

For instance, red coat color in Angus cattle is a recessive gene, which by the previous discussion of gene dominance, would indicate that 25% of all Angus cattle would be red. This is not true, because there is not an even distribution of genes. There is actually a much greater percentage of dominant (black) genes, which means that most Angus cattle are black.

Another example would be the presence of horns in cattle. Horns are controlled by a recessive gene. In some breeds of cattle the gene frequency for horns is very high, and as a result, the incidence of horns is very high. When crossed with cattle that are polled (carry the dominant gene), the offspring will always be polled. This is why black baldy calves from Hereford x Angus matings are always polled.

ADDITIVE GENE ACTION

As mentioned earlier, qualitive traits, or those traits which can be seen (horns, coat color, etc.) are controlled by only one pair of genes. Quantitative traits, or those traits which must be measured (weaning weight, feedlot gain, etc.), are controlled by large numbers of genes. These genes are said to have an additive effect; i.e. collectively the genes produce the trait. Thus, if the calf receives a large number of "good" genes for weaning weight, it will be a heavy calf. If the calf receives only a few "good" genes, it will be somewhat smaller.

HERITABILITY

Heritability is defined as a measure of the ability of an animal to pass on quanitative traits to its offspring; i.e. those traits controlled by additive gene action. Expressed as a decimal, it is a measure of what can be expected from the offspring of parents with known records for a quanitative trait.

For example, the heritability for weaning weight is generally accepted to be .3. In a herd that has an average weaning weight of 400 lb, this means that if a heifer were selected that weighed 450 lb at weaning . . . as a cow* she would have calves that would wean 7½ lb heavier than the average; i.e.

450 lb	heifer wean weight
-400 lb	herd average
50 lb	differential
x .3	heritability
15.0 lb	
x .5	amount of genes contributed by the cow
7.5 lb	increase in weaning wt. over herd ave.

Likewise if the heifer were bred to a bull that weaned 50 lb heavier than the other herd bulls, another 7.5 lb increase in weaning weight could be expected.

Justification for the existence of the artificial insemination industry can quickly be recognized when one contemplates the advantage to breeding to a bull with a weaning weight of 200 or more pounds greater than the herd average for bulls. Calves

*First and second calf heifers typically wean smaller calves than mature cows.

with weaning weights 30 lb greater than the herd average could then be expected just from the bull's genes; i.e.

200 lb	differential
x.3	heritability
60 lb	
x.5	amount of genes supplied by bull
30 lb	increase in weaning weight over herd average

Obviously bulls of this quality would be very expensive . . . the cost of which could not be justified to breed 20 or 30 cows a year. Thus, artificial insemination has become a viable business.

HYBRID VIGOR OR HETEROSIS

Hybrid vigor or heterosis is defined as the phenomenon associated with the crossing of two breeds or species of animals. For a yet unexplained reason, hybrid offspring will usually be superior in traits that are otherwise lowly heritable. Fertility is one of the most important traits with a low heritability, and crossbred cows are usually more fertile than either parent breed. Likewise, crossbred calves are typically much hardier and therefore have a higher survival rate.

The amount of heterosis to result from crossbreeding is generally accepted to be inversely proportional to the heritability of the trait in question. This explains the increase in fertility and hardiness. Quite often, however, increases in relatively heritable traits such as weaning weight and feedlot gain are also experienced (Table 13-1). This can be explained by better adaptation, and resistance to disease, as well as simple genetic up-

Table 13-1.   WEANING   WEIGHT   OF   COMMERCIAL   ANGUS, HEREFORD, AND ANGUS X HEREFORD CROSSES.*

Breeding of Calf	Birthweight	Weaning Wt.
Angus X Angus	68.8 lb	393.4 lb
Hereford X Hereford	73.0 lb	383.5 lb
Angus X Hereford	71.7 lb	407.7 lb
Hereford X Angus	70.2 lb	406.3 lb

*Adapted from Gray (1978)

grading; i.e. oftentimes straightbred commercial herds are some-what inbred, or of otherwise lowly genetic makeup. When moving into a crossbreeding program, many operators become more conscious of genetic quality and will buy better quality bulls than they did previously.

GENETIC ASPECTS OF ANIMAL BREEDING

CROSSBREEDING
Crossbreeding may be defined as the breeding of different cattle breeds so as to take advantage of heterosis in the offspring. Crossbreeding has found its greatest acceptance and use in areas which create environmental stresses, heat stress in particular.

Defined as hybrid vigor, crossbred animals are generally hardier and more able to withstand harsh environments. The classic example of this was the development of the Santa Gertrudis breed at the King Ranch. Situated along the very hot and humid Texas Gulf Coast, the environment greatly hindered the production of Shorthorn cattle on the ranch. Brahma cattle were also present, but while they are able to withstand tropical environments, Brahma cattle are not particularly productive, and are relatively infertile. Crossing the Brahma cattle with the Shorthorn resulted in a breed that was much more fertile and productive than either of the two parent breeds. Since then there have been a number of crossings between Brahma and European breeds to produce cattle better adapted to the southern and southwestern U.S., as well as Central and South America; e.g. Brangus, Beefmaster, Barzona. Braford, and more recently, Bramental.

In the more temperate regions, crossbreeding is practiced because it usually produces a greater total amount of calf weight weaned. Weaning weights are not only greater, but also more calves usually survive to be weaned. Crossbred calves seem to be hardier and have a greater survivability as baby calves than calves from straightbred cows.

Selecting the breeds to be crossed will depend upon several factors, the most important being the environment. As a general rule, the more diverse the breeds, the greater the response to crossbreeding will be. For example, a greater response would be obtained by crossing the English breeds (Hereford, Angus, Shorthorn) with one of the Continental breeds such as Charolois, than if they were crossed among themselves.

If crossbred heifers are used as herd replacements, weaning weights are usually improved further as crossbred cows typically give more milk. Likewise, crossbred cows are usually more fertile, and so greater overall calf crops are usually obtained (in addition to increased survivability of the calves themselves) (see Table 13-2).

Table 13-2. CALF SURVIVABILITY AND WEANING WEIGHT COM— POSIT DATA FROM UNIVERSITY RESEARCH REPORTS. Adapted from Long, 1980.

Breed	Calving rate, %	Calf survival to weaning, %	Weaning weight lb	Source of data
Brahman (B)	81.3	92.1	397.5	Texas: 585 matings.
Hereford (H)	80.1	90.0	390.0	Cartwright et al.
B X H and H X B	87.2	91.0	427.9	(1964).
Angus	65.1	92.6	411.4	Louisiana: 1,345
Brahman	67.9	92.0	416.9	matings. Turner
Brangus	65.6	92.4	438.2	et al. (1968),
Hereford	65.6	91.0	389.0	McDonald & Turner
Angus X Brahman	79.7	93.7	445.7	(1972).
Brahman X Angus	77.4	80.5	469.5	
Angus X Brangus	63.5	97.9	434.7	
Brangus X Angus	81.8	92.6	432.3	
Angus X Hereford	68.2	90.7	427.2	
Hereford X Angus	71.4	97.5	432.1	
Brahman X Brangus	82.1	89.9	455.6	
Brangus X Brahman	68.8	95.5	444.8	
Brahman X Hereford	84.7	96.0	454.5	
Hereford X Brahman	86.4	96.4	471.2	
Brangus X Hereford	77.1	94.6	422.6	
Hereford X Brangus	67.1	91.8	439.6	
Brahman (B)	71	97	396	Florida: 1,135 matings.
Shorthorn (S)	64	94	334	Peacock et al. (1971),
B X S and S X B	76	96	429	Kroger et al. (1975).
Angus	83.2	91.6	440.9	Nebraska: 1,257
Hereford	83.5	93.7	419.3	matings. Cundiff
Shorthorn	76.8	93.6	440.9	et al. (1974).
Angus X Hereford	96.8	93.2	454.5	
Hereford X Angus	89.1	89.0	426.6	
Angus X Shorthorn	85.9	96.0	464.9	
Shorthorn X Angus	84.2	93.4	457.6	
Hereford X Shorthorn	80.0	96.8	454.3	
Shorthorn X Hereford	87.3	94.7	455.2	
Angus	84.2	84.1	457.8	Indiana: 693 matings.
Milking Shorthorn	76.5	85.0	482.9	Spelbring et al.
Angus X M. Shorthorn	87.6	92.5	484.4	(1977).
M. Shorthorn X Angus	86.5	90.1	486.9	
Angus	96.3	97.1	449	Virginia: 604
Hereford	92.1	95.8	420	matings. Gaines
Shorthorn	89.9	91.0	438	et al. (1978).
Angus X Hereford	97.9	95.6	471	
Hereford X Angus	94.3	94.1	442	
Angus X Shorthorn	98.0	87.8	466	
Shorthorn X Angus	92.3	87.5	480	
Hereford X Shorthorn	91.2	96.2	422	
Shorthorn X Hereford	97.2	90.3	464	

OUTBREEDING

Outbreeding simply refers to the breeding of unrelated individuals. While crossbreeding is technically a form of outbreeding, the term usually refers to matings within a breed.

INBREEDING

Inbreeding refers to the mating of related animals. Under most conditions inbreeding should be avoided as it will increase the incidences of undesirable traits such as dwarfism. These traits are usually recessive homozygous, and inbreeding increases the probability of bringing out the recessive traits.

Traits controlled by additive genes are also adversely affected as weaning weight and feedlot gain are reduced in inbred cattle. Scientific estimates of reduced weaning weight due to inbreeding are approximately 1 lb for every percent of inbreeding (Pollak, 1978; Brinks, 1975).

Inbreeding has a place in purebred breeding as a means of proving or disproving that an animal (bull in particular) is or is not a carrier of an undesirable gene. By mating to close relatives, the probability of the gene expressing itself will be much greater.

LINEBREEDING

Linebreeding is a form of inbreeding in which distant relatives are bred to an outstanding relative. This practice has a place in purebred breeding as it tends to "fix" the genes and thus make the animals more homozygotic. As seed stock then, the animals are more prepotent or true breeders.

This practice should only be undertaken with caution. The traits for which the genes are to be "fixed" should be examined carefully, and the ancestories of the individuals researched thoroughly. The breeder should be convinced that the desired traits to be fixed will indeed be accomplished by the matings. If the linebreeding is to be done simply because the breeder feels the one individual is outstanding, the decision should be based on statistical evidence. Oftentimes visual appraisal and even personal prejudice have been the only criteria used for entering into linebreeding programs. This type of decision making only leads to reduced weaning weights, and calf survival.

NICKING

"Nicking" is a term used (typically by purebred breeders) to denote matings of 2 individuals or lines of cattle which produce offspring that are superior to what would ordinarily be expected from parents of that caliber. That is, matings between certain individuals or lines sometimes result in more favorable gene combinations than if those same animals were mated with other individuals or lines with similar production records. Thus, it is said that certain bulls will nick with certain lines of cows.

The phenomenon is not completely understood, but it is thought to be the result of epistasis, or interaction between otherwise unrelated genes. For example, it is known that in laboratory animals and some vegetables that genes for different traits can create third traits when brought together (Rice, 1970). It is therefore theorized that additive genes in cattle can do the same thing (produce superior performance) when brought together in unknown combinations.

SELECTION INTENSITY

Selection intensity is a term used by geneticists in predicting response or progress. Selection intensity is a function of selection differential. In layman's terms, selection differential means the differences between the herd average and new animals introduced into the breeding herd, or conversely, the differences before and after culling. The greater the difference, the greater the response will be. In other words, if bulls that were 100 lb heavier at weaning (than the herd ave.) are selected over bulls that were only 50 lb heavier, then weaning weight of the calf crop will obviously be somewhat heavier. Selection intensity would refer to the number of higher quality bulls used as replacements, or the total number of cows culled. Obviously, the greater the number of quality bulls, and/or inferior cows culled, the higher the genetic progress will be.

Selection intensity (and progress) is also affected by the number of traits animals are selected for. If more than one trait is selected for, then progress for any one trait will be reduced. This concept is of vital importance to the beef industry, and in the past has been the result of some serious genetic setbacks. If it is increased weaning weight that is desired, then it is increased weaning weight that must be selected for. If we want calving ease as well as weaning weight, then obviously we cannot expect maximum gains for either trait since growth is positively correlated with birthweight. This is not to say that one should not select for both weaning weight and calving ease as there are situations where these two traits should be considered together (page 273 & 289). We should simply be cognizant that movement toward either trait will retard progress for the other trait.

However, the author would like to point out that in the past there has been selection for traits that were purely cosmetic. As mentioned on page 288, this has set back the growth rates on many lines of cattle. Clearly, if we want cattle that are efficient producers of meat, we must not be concerned with coat colors, presence or absence of spots or markings, etc. Selection must be based on only those traits that make cattle more productive.

CHAPTER 14   REPRODUCTIVE PHYSIOLOGY

An understanding of the reproductive functions in cattle is of obvious importance for a cow-calf operator. A knowledge of the female is far more important to the producer, since the technical problems he must deal with primarily concern cows and heifers; e.g. dystocia (difficult birth), recognition of estrus (cows being settled), artificial insemination, etc. For that reason, the main emphasis in this chapter will be on female reproductive anatomy and physiology, and the common problems and management practices the producer will be involved with.

REPRODUCTIVE ANATOMY OF THE BULL

The reproductive anatomy of the bull is quite similar to most mammals. Sperm cells are produced in the testes, and stored in the epididymis (tubular like storage vessel attached to the testes). The epididymis is connected to the urethra by a tube known as the vas deferens.* Just before reaching the urethra, the vas deferens opens up into a much larger tube, known as the ampulla. During sexual excitement sperm cells are transferred from the epididymis to the ampulla. Simultaneously the prostate and the bulbourethral glands secrete accessory fluid to cleanse and lubricate the urethra. When ejaculation occurs the vesicular glands secrete accessory fluid to add volume to the semen (semen is considered to be the sperm cells plus the accessory fluids).

The penis of the bull is what is known as the fibroelastic type. This type of penis does not engorge with blood, but remains relatively rigid all of the time. During copulation the sigmoidal flexure (Figure 14-1) is straightened out, which allows the penis to be driven out of the sheath to penetrate the female.

The amount of sperm cells produced in the testes will vary with the amount of sexual activity. If the bull has been sexually inactive for a period of time, sperm production will be greatly reduced. If the period of inactivity is extensive, sperm viability will deteriorate. Because of this, the first ejaculate of the breeding season will often produce very poor quality semen; i.e. sperm cell motility will be greatly reduced. Normal sperm are very active "swimmers", but sperm cells from the first ejaculate will usually be sluggish or totally inactive. Fertility testing of bulls involves

---

*In a vasectomy, the vas deferens is severed so that sperm cells cannot reach the urethra. Vasectomized bulls are sometimes used for heat detection in artificial insemination programs.

Figure 14-1.　REPRODUCTIVE TRACT OF A BULL.

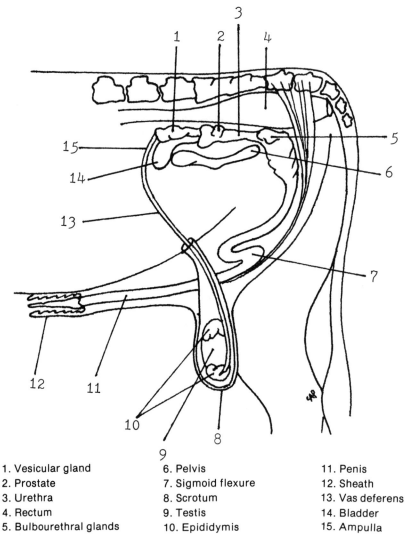

1. Vesicular gland	6. Pelvis	11. Penis
2. Prostate	7. Sigmoid flexure	12. Sheath
3. Urethra	8. Scrotum	13. Vas deferens
4. Rectum	9. Testis	14. Bladder
5. Bulbourethral glands	10. Epididymis	15. Ampulla

viewing sperm cells under a microscope, and motility is one of the most important criteria. Obviously, if bulls are to be fertility tested at the beginning of the breeding season, two ejaculations (some time apart) may be necessary. Likewise, if hand mating is practiced, a second service should be allowed.

An enormous amount of sperm is normally contained in an ejaculation. While the actual numbers will vary with the individual bull and previous sexual activity, a range of 5 to 15 billion can be anticipated. (It should be pointed out that with some bulls the lower range can be "0".) When semen is deposited directly into the uterus (as in artificial insemination), only about 5 to 10 million sperm cells are required. Thus, through artificial insemination, one bull can be used to breed thousands of cows.

REPRODUCTIVE TRACT OF THE COW

The reproductive tract of the cow is relatively similar to that found in other farm animals. The ovaries, or female gonads, are attached on both sides of the body cavity, just ahead of the pelvis, above and behind the uterus. The ovaries produce eggs or ova which are carried down the oviduct to the uterus.

Figure 14-2. REPRODUCTIVE TRACT OF THE COW.

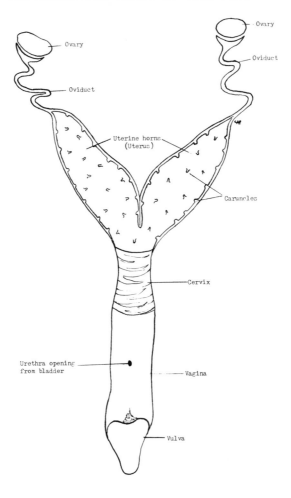

If sperm cells are present and fertilization occurs, the fertilized ovum (embryo) will develop in the uterus. This can take place in either horn (side) of the uterus. The caruncles, small protrusions inside the uterus, are used to transfer nutrients from the uterus to the placental tissues during pregnancy.

The cervix, which is located between the vagina and the uterus, plays an immensely important role in reproduction. Resembling and feeling like a turkey neck, it is composed of annular rings of cartilaginous tissue. During pregnancy the cervix closes tightly and secretes a mucous plug to effectively seal off the uterus from outside bacterial contamination (Figure 14-3). During estrus, the cervix opens to allow the passage of sperm cells, and completely dilates during parturition to allow passage of the fetus.

Figure 14-3. CERVIX SEALING OFF UTERUS DURING PREGNANCY.

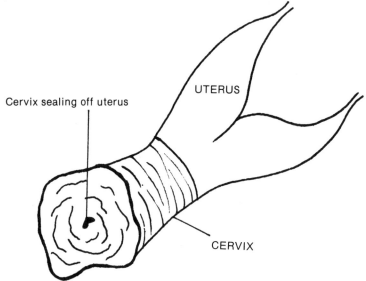

Attached to the posterior of the cervix is the vagina, which is a tube of elastic tissue which receives the bull's penis and semen during copulation. During parturition the vagina greatly dilates to afford passage of the fetus. Below the vagina lies the bladder which attaches to the vagina via the urethra at about the midpoint of the ventral (bottom) side of the vagina. During urination, the urine travels through the posterior half of the vagina to the vulva, which is the external female genitalia.

THE OVARIAN CYCLE

Mature nonpregnant cows go through a regular ovarian cycle

which normally takes an average of 21 days to complete. Under-standing this cycle is a basic requirement to those engaged in cow-calf production.

The cycle begins with the secretion of gonadotropic hormones from the pituitary gland, which initiates the production of ova or eggs in the ovaries. The ova develop in a fluid-filled compart-ment known as a follicle (Figure 14-4).

Figure 14-4. FOLLICLE ON OVARY.

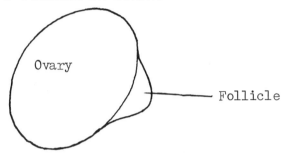

Estrogen, a female hormone, is produced within the follicle. Just prior to rupture of the follicle, large amounts of the estrogen are released into the bloodstream which causes the cow to display signs of being in estrus or heat.

During estrus, the cervix will dilate to allow passage of sperm, and mucus will be secreted in both the cervix and the vagina to facilitate copulation. A pheromone (hormone like odor) is given off, which will attract bulls and "tell" them that the cow will stand for a service. The cow will stay in a "standing heat" for approxi-mately 12 to 18 hours.

A few hours after the cow passes out of heat, the follicle will rupture and the ova will pass down the oviduct. If the cow has been served, sperm cells will usually be present in the oviduct and fertilization will occur. After a period of a few days the ova will pass through the oviduct into the uterus and, if fertilized, will attach itself to the uterine wall and a fetus will develop.

Regardless of whether or not the egg has been fertilized, the ruptured follicle on the ovary will regress to form a glandular structure known as a corpus luteum. The corpus luteum will then secrete a hormone known as progesterone which will cause the cervix to close, and otherwise reverse all signs of heat. Under the influence of progesterone the cow will become calm and will refuse to be serviced.

If fertilization occurred, the corpus luteum will remain and continue to secrete progesterone until the time of parturition. If fertilization did not occur, the corpus luteum will recede in a few days, and new follicles will develop.

This entire cycle takes place about every three weeks, and therefore open cows will come into heat about every 21 days. The actual times may vary, but a maximum variation in cows that are reproductively sound would be ± 7 days.

METHODS OF MATING

There are three methods of mating used for beef cattle; hand mating, pasture mating, and artificial insemination.

HAND MATING

Hand mating consists of keeping the bull separate from the cows to be bred. When a cow is found to be in heat, she is brought to the bull and serviced. This method of mating is practiced primarily by purebred breeders who need to be certain of the parentage of each calf. Obviously this method requires a considerable amount of labor.

An advantage to this form of mating over pasture mating, is that one bull can be used to serve many more cows. In a pasture breeding situation a bull will often breed a cow several times during her heat period. Repeated breeding will, of course, reduce the number of mature sperm cells in the ejaculate. Therefore, fewer cows can be served.

PASTURE MATING

Pasture mating is the form of mating used by the majority of commercial operations. The advantages are the obvious reduction in required labor. After the bulls are put in with the cows, all that is required are occasional riders to see that the bulls are actually with the cows, and that the cows are actually being settled. While the term "occasional rider" may seem casual, the responsibility should not be taken lightly. An experienced, responsible man should be given this job, as failure to be observant could lead to financial disaster for the ranch. Most anyone can tell if the bulls are actually following the cows, but it takes an observant man to notice if the cows are actually being settled.

If toward the end of the breeding season it is noted that cows are coming back into heat, the situation must be investigated promptly. Much has been said concerning fertility testing of bulls, and certainly infertility in one or more bulls could cause such a problem. More commonly however, nutritional deficiencies are the culprit (page 274). The worst possible situation would be a venereal disease outbreak. But regardless of the cause, the situation must be detected early.

The number of cows a bull can service under pasture breeding will depend greatly upon the types of pastures. In irrigated or improved pastures where there are no physical obstructions (mountains, deep ravines, dense brush, etc.) and the cows are concentrated together, a mature bull could be expected to service up to a maximum of 40 cows. Under range conditions, 20 to 25 cows per bull is much more common. Where forage is extremely sparse and the stocking rate very light, ratios of 15:1 may be appropriate.

In each ratio given it is assumed that the bulls are three years or over in age. Two year olds should be turned out only if they have proven themselves to be aggressive breeders. Even then it is best to reduce the number of cows they are to serve. It is also best to place two year olds with the replacement heifers, as sometimes they are hesitant to breed mature cows, or may not be tall enough to effect a satisfactory penetration; i.e. up to the cervix. Likewise it is best not to put two year olds with mature bulls as they are often intimidated by the older bulls. Yearling bulls cannot be depended upon to successfully conduct pasture mating.

ARTIFICIAL INSEMINATION

Artificial insemination (A.I.) has long been practiced in dairy breeding. Dairying, of course, lends itself to A.I. better than beef breeding, since the cows are confined and otherwise handled daily. But probably more than that, the very nature of dairying requires record keeping, and astute dairymen have long since learned that better milking heifers can be produced from bought semen than from the bulls they can afford to buy. The use of genetically proven A.I. sires has been a primary factor in the dairy industry producing the same total quantity of milk using half the dairy cows, in less than 20 years (Beverly, 1978).

Due to the physical problems involved, however, A.I. has never been particularly popular with beef breeders. During the early 1970's when the industry went through the so called exotic or Continental European beef breed fad, A.I. experienced a mild boom; i.e. due to hoof and mouth import restrictions, the only feasible way for a stockman to obtain European cattle was through the purchase of semen. The excitement created by the "exotics" has waned, and consequently the use of A.I. on beef cattle has also receded. However, the use of A.I. definitely has a place in beef cattle breeding. Not only does it give the operator access to bulls of genetic worth that he could never afford to purchase, but it also greatly facilitates crossbreeding programs, and eliminates diseases spread by venereal contact. On the negative side, it does require more labor and a greater sophistication of management.

The Technique In General - The purpose of this section is to describe the basic techniques used in artificial breeding. The idea being to demonstrate to the reader that the practice is relatively simple. Detailed instructions are best provided by the commercial bull stud providing the semen. Likewise, the major bull studs and some universities regularly conduct short courses in A.I. which are designed to make the operator sufficiently competent to breed his own cows.

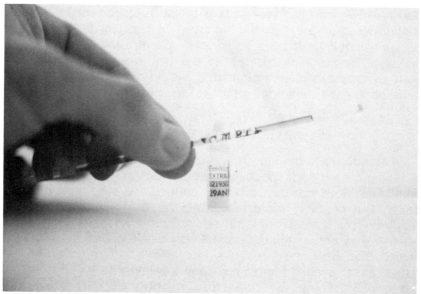

Figure 14-5. Ampule and straw of semen.

The semen that the stockman buys comes to him in either an ampule or straw, frozen and stored in liquid nitrogen. One ejaculation can be used to breed hundreds of cows, and so the ampule or straw contains only a fraction of an ejaculate mixed with an extender. This is why semen from bulls that would cost enormous sums of money, can be purchased for nominal fees.

The semen is delivered and kept in an insulated tank containing liquid nitrogen. The first step in A.I. (after finding a cow in heat) is to remove the semen from the tank and thaw it. The company merchandizing the semen will provide detailed instructions on how to thaw the semen, and these instructions should be followed closely as the extenders in the semen may vary with the processor. Generally though, ampules are placed in ice water, and straws allowed to thaw in the air (after placement in an insemination catheter).

After thawing, ampules are cut and/or broken and the semen siphoned up into the breeding catheter. This is either accomplished by a small rubber bulb (poly bulb), or a small syringe attached by a short piece of flexible tubing (Figure 14-6).

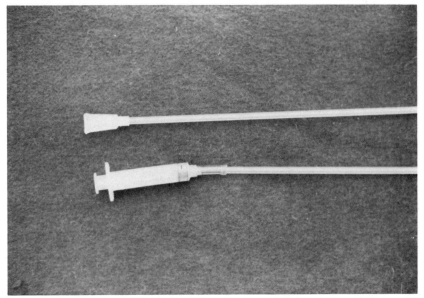

Figure 14-6. Inseminating pipette with poly bulb (above) and with syringe (below).

The cow to be bred is placed in a simple chute (Figure 14-7), with a pipe or post placed low behind her hind legs (to reduce danger from kicking, as well as to provide restraint). Squeeze chutes can be used, but tend to excite the animals.

The breeder then places an obstetrical sleeve on one arm, lubricates it (soapy water works quite well), and places it in the rectum of the cow. This is best done by forming a wedge with the fingers and slowly pushing the hand in as the anal sphincter relaxes.

Deposition of the semen should be made deep within the cervix, near the uterus, and so after insertion of the hand into the rectum, the operator must find and "pick up" the cervix (through the rectal

Figure 14-7. IDEAL PALPATION CHUTE.

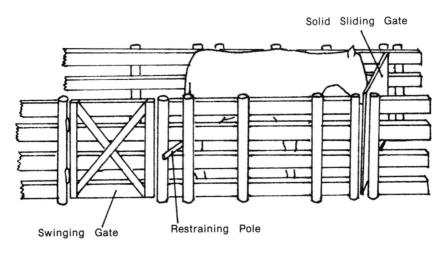

Solid Sliding Gate

Swinging Gate

Restraining Pole

wall). The cervix is the most distinctive organ to be found in the
pelvic region as it has the approximate size and feel of a turkey
neck (see Reproductive Tract of the Cow, Figure 14-2). The cervix
is attached to the anterior (forward) part of the vagina and is usually
found "lying" on the floor of the pelvis (Figure 14-8).

Figure 14-8. FEMALE REPRODUCTIVE TRACT AS VIEWED FROM
ABOVE. (As would be felt during palpation or artificial insemination)

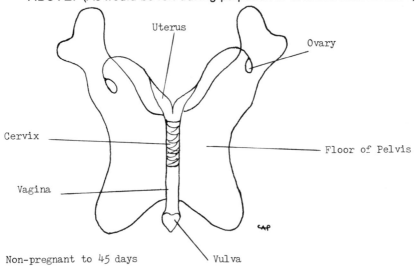

Uterus

Ovary

Cervix

Floor of Pelvis

Vagina

Non-pregnant to 45 days

Vulva

Note: As pregnancy develops cervix will move forward and drop over the brink
of the pelvis.

When the cervix is located, the catheter is then placed into the vagina. Absolute cleanliness is essential, as bacteria introduced into the vagina or cervix can create infection. Most technicians will exert a downward pressure with the arm that is in the rectum, which causes the lips of the vulva to "pop" open. The catheter can then be placed into the vagina without coming into contact with contamination on the vulva.

The catheter is then guided into the cervix with the hand holding the cervix from the rectum above. When the tip of the catheter is up next to the uterus, the semen is deposited (Figure 14-9).

Figure 14-9. PROPER SITE OF INSEMINATION.

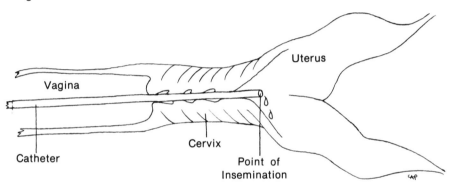

The walls of the vagina and cervix are fairly tough, but the uterus can be punctured quite easily. The operator should therefore use extreme caution when approaching the end of the cervix. The catheter should never be pushed or forced into the animal. During estrus the vagina and cervix will be well lubricated with mucus so the catheter should slide easily. The difficult part is finding the orifice of the cervix; the vagina has a diameter of one to one and a half inches, but the cervix has only about a ¼ inch opening (Figure 14-10). When the semen is actually deposited, the plunger should be pushed or squeezed gently so as to minimize physical damage to the sperm cells.

When To Physically A.I. A Cow In Heat - As mentioned in the section on hand mating, ovulation usually occurs after the cow passes out of "standing heat". Sperm cells are generally thought to remain viable for 12-24 hours, and since cows can remain in heat up to 18 hours, many authorities recommend breeding several hours after the cow is discovered to be "in". If the stockman does his own work, this is probably good advice. American Breeder's

Figure 14-10. LONGITUDINAL CROSS SECTION OF VAGINA
AND CERVIX.

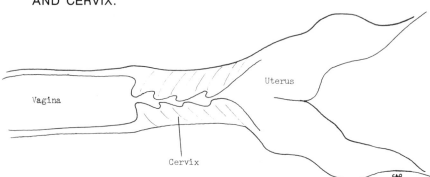

Note: Notice the blind endings in the cervix which can make passing an inseminating catheter difficult.

Service (ABS), the largest A.I. organization in the world has
researched this however, and has found very little difference
in the conception rate between cows bred when discovered, and
those given a delayed breeding. For this reason, ABS recommends
breeding as soon as possible, to avoid any chance of the cow not
being inseminated due to the breeder being called away, detained,
etc.

Heat Detection - The most difficult part of any artificial insemination
program is the process of detecting heat in the cows to be bred;
difficult in that it requires the most time and labor. Throughout
the breeding season, the cows must be observed carefully for
1 to 3 hours every morning and evening. The morning observation
should begin just before daylight, and the evening observation
should end just before dusk. Failure to do a proper job of heat
detection is the most common cause of A.I. program failure. The
commercial bull studs typically do everything possible to help
their stockmen/customers set up a workable heat detection program, but the end result is up to the determination of the operator.

Signs Of Heat - The most well known, and the most certain sign
of estrus is the "standing heat". Cows or heifers in heat give
off a pheromone (odor) that alerts the rest of the herd as to their
condition. Other cows or heifers will then come up and mount
them as a bull would do. Females that are in heat will "stand
to be ridden"; females that are not in heat will not "stand". It
is the heifer on the bottom . . . the one that stands, that is in
heat. Occasionally a female in heat will ride others, but when

approached by others, she will also usually stand to be ridden herself.

Other signs of estrus would include bellowing, walking (pacing) the fence (separation from the herd), going off feed (not grazing while others are grazing), or otherwise unusual behavior. The tail head may be ruffed up from being ridden by other females, and sometimes mucus can be seen on the vulva.

Estrus will generally be expressed early in the morning or late in the evening and so this is when observation should take place. A very effective technique is to herd the cows together and let them sniff and inspect each other. Stirring them around will usually stimulate riding activity.

AIDS IN HEAT DETECTION

Heat Sensitive Patches - In addition to visual observation there are a number of devices and practices used to detect heat. Pressure sensitive patches (trade name Kamar) were one of the first items to be used, and are reasonably effective in certain situations. The patches contain a tube which under pressure will change colors (white to red). The patches are glued to the tailhead of the cow and when others ride her, the pressure exerted from the sternum of the mounting animal causes the color change. The main drawback to the patches is that if trees or brush are in the pasture, the cows will often rub the patches against limbs, etc., and activate them.

Sterilized Bulls - Bulls sterilized by different means are often used as an aid. Penectomized bulls and the use of a devise that prevents the penis from leaving the sheath have been used, although libido (sex drive) sometimes decreases. Vasectomized bulls have been very popular. The bulls are fitted with a chin ball marker, or sometimes paint on the brisket. Thus when the bulls mount a cow, a mark is left on the back or tail head which indicates that the cow is in heat. While the use of sterilized bulls does greatly facilitate heat detection, it obviously creates additional expense. The bull expense is not as great as with natural service, however, since quality bulls are not required. The use of vasectomized bulls does not eliminate the potential for venereal disease, which is otherwise an advantage to A.I.

Androgenized Cows - A technique for heat detection known as androgenized cows came into commercial use about 1978. An androgenized cow is one that has been injected with synthetic male hormone (testosterone compounds) to create male-like behav-

Figure 14-11. Sterilized bull fitted with a chin ball marker as an aid in heat detection (courtesy American Breeder's Service).

ior. Cows treated in this manner will seek out and mount other cows in heat. At the time of writing the technique was fairly new, but American Breeder's Service (1979) reported successful use in 94 different herds using 419 androgenized cows.

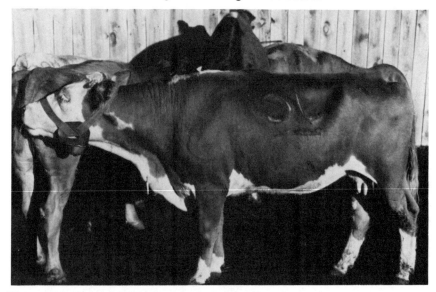

Figure 14-12. Androgenized cow fitted with chin ball marker for heat detection (courtesy American Breeder's Service).

Treatment of the cows is rather simple logistically, but very specific as to the type and concentration of the hormone used (testosterone enanthate and testosterone propionate were the compounds being used at the time of writing). For detailed instructions the producer should contact a reputable breeding service or knowledgeable veterinarian.

The primary advantage to androgenized cows (over sterilized bulls) is ease of preparation. There is no surgery, no recovery period, and little chance of infection. The other advantage is that cows are readily available from the herd (cull cows). There is no need to bring in outside bulls and thus take a chance on introducing disease into the herd (or the expense of keeping bulls year round). After the breeding season (and a 14 day withdrawal period), the cows may be sold for slaughter.

ESTRUS SYNCHRONIZATION

At the time of writing, the FDA had just cleared the use of Prostaglandin, a hormone like drug, as a means of estrus synchronization. When injected into cows and heifers, Prostaglandin causes the corpus luteum to recede. As explained earlier the corpus luteum forms on the ovary after an egg is released, and secretes progesterone which keeps the cow or heifer from coming into heat. The corpus luteum functions for about 18 days, and then recedes. In about three days a new follicle will develop and secrete estrogen, which causes the animal to go into heat. Prostaglandin can cause the corpus luteum to recede much more quickly. If a cow or heifer is between day 6 and day 17 of her ovarian cycle (cycle normally runs for 21 days), injection with Prostaglandin will cause the animal to go into heat in 2 to 5 days. Cows between day 1 and day 6, and day 18 and day 21 are not affected by Prostaglandin. If pregnant cows are injected, abortion may result (an FDA clearance is being sought as a feedlot heifer abortant).

The obvious advantage to Prostaglandin is that it shortens the insemination period. At the same time, however, it greatly intensifies the labor requirement for that period of time. Heat detection is required in most management programs utilizing Prostaglandin, and can become difficult with so many animals coming into heat at once.

PREGNANCY TESTING

Pregnancy testing by rectal palpation has become a fairly common and useful management tool. The practice is to test the cow herd 3 to 4 months after the breeding season; i.e. just before winter in the spring calving operations and just before summer in fall calving operations. By testing just before the period when supplemental feeding is necessary, costs are reduced by culling barren cows before $50 or $75 in purchased feed is invested in them. Then too, in a spring calving operation, cows will be heavier before winter than they would be afterward. Since cull cows are sold by the pound, it makes no sense to sell a 900 lb cow in the spring, when she probably weighed close to 1,000 lb in the fall (without the supplemental feed investment).

THE RECTAL PALPATION TECHNIQUE

The reproductive tract of the cow lies directly under the rectum and the entire tract can usually be palpated (felt), by inserting the arm in the rectum. Veterinarians are most often called upon to do the work, but the technique requires only a basic knowledge of anatomy (the whereabouts of the reproductive tract), and so most stockmen can learn to test their own cattle. Moreover, proficiency comes primarily as a result of practice and experience.

The best facility for pregnancy testing is a single file wooden chute such as the kind that lead up to squeeze chutes. While squeeze chutes themselves can be used, they usually tend to excite the cattle. For rectal palpation it is best to have the cows as calm as possible. Wooden chutes that are narrow enough to prevent the cow from turning around are best. Restraint in front of the cow should be in the form of a solid gate; a sliding gate is ideal. Pipes or poles placed behind the cow are quite satisfactory, but should not be used in front, since if cattle can see ahead, they will try to jump; i.e. solid gates block their view. The pole placed in the rear should be placed low enough that if a cow tries to lie down, the palpator can pull his arm out of the rectum before it comes in contact with the pole. Likewise the low pole serves as an aid in guarding against kicking (see Figure 14-7).

Plastic obstetrical sleeves are commonly used to protect the arm as they are cheap and disposable. In learning situations where a cow is to be palpated for an extended period of time, the sleeve should be turned inside out so that the seams do not irritate the intestinal mucosa. Prior to actual use it must be lubricated. Mild soap and water makes excellent lubricant. Strong soaps and detergents should not be used as they will irritate the cow.

The rectal membrane of the cow is reasonably tough, but it can and has been punctured during palpation so care should be exercised. When entering a cow, the fingers and thumb should be held together in the shape of a wedge. Gently force the hand and arm into the rectum. If the cow tries to expel your arm simply "hold your ground"; don't try to force the hand any farther until the contractions relax. Once inside, use the hand like a paddle, and feel with the entire hand rather than the individual fingers. To move deeper into the cow, extend the arm only gently and slowly. Using these precautions, the operator should never have any problems. Quite often small blood vessels are ruptured which will cause a slight amount of bleeding. This usually causes no problem. If the intestinal lining begins to feel rough (normally it feels slimy), this means the mucosal lining has been worn off. Without the intestinal mucosa there is a much greater chance of puncturing the cow, and so the arm should be withdrawn.

The first organ to look for is the cervix. As mentioned earlier, it is the most distinctive organ in the reproductive tract as it has the gristly consistency and feel of a turkey neck. In "open" cows or cows less than 2½ months pregnant, the cervix should be on top of the pelvis, although it may be off to the side. Once the cervix is located (see Figure 14-8), the operator can follow it forward to find the uterus. In pregnancies of 30 days or less there will be no distinguishable fetus, only a swelling in the uterine horn containing the pregnancy. At 45 days there will only be a fluid filled sac (amnion) about the size of a marble. By 2 months the entire uterus will be filled with fluid and a fetus about the size of a mouse will be present. Detecting pregnancies prior to two months is extremely difficult for someone with limited experience.

Figure 14-13. PREGNANCY DETERMINATION BY PALPATION.

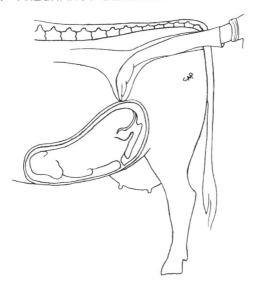

After three months the cervix will be pulled over the brink of the pelvis. Not being able to find the cervix in itself is a reasonable indication of pregnancy.

From two to four months the fetus can be palpated by slowly patting the area in front of the pelvis. When the fetus comes in contact with the hand, it will "bounce" down, and then come back up and bounce against the hand a second time. It is the only solid object that will feel suspended. In some cases the fetus may be located well over the brink of the pelvis, which may necessitate pushing the arm in all the way to the shoulder. When this is the case, feel down and all around, as the reproductive tract is only loosely attached and therefore can be most "anywhere".

At five months the fetus will be the size of a puppy. It will be way down, and touching it may push it completely out of reach. By six months the limbs will have begun to grow. The palpator will probably touch a foot first as normal carriage at this time is with the feet sticking up. From seven months on, the fetus will be quite large and so palpation will be extremely simple.

PARTURITION

INDICATIONS OF PARTURITION

Cattle are said to have a gestation length of 283 days, but obviously that will vary between individual and breeds. Outside of gestation length, there are observations which are indicators of approaching parturition.

As the cow moves into the ninth month of pregnancy, the vulva has a tendency to "pull in and pop out" as the cow walks. This is caused by the weight of the fetus pulling on the reproductive tract. Likewise, the abdominal section of the cow will appear quite large and "heavy with calf". In western U.S. vernacular, an obviously pregnant cow is said to be a "springing cow". One should be cautious however, as a heavy fill of wet feed such as lush grass or silage can greatly extend the abdomen. Likewise, giving cows sudden access to good quality hay such as alfalfa, and clean water can also produce a "fill" that can be mistaken for pregnancy. However, the bobbing in and out of the vulva is fairly distinctive.

From 1 to 3 weeks before calving, the udder will begin to fill and the teats will appear turgid. (In feedlot heifers this is not an indication as hormonal implants often cause the same effect.) When parturition is imminent, the cow will isolate herself from the herd, lie down and get up repeatedly, kick at her belly, glance to the rear nervously, and otherwise appear uneasy. When labor actually begins, rhythmic abdominal contractions will occur.

BIRTH OF THE CALF

Normal position of the calf at the time of birth is with the front feet and head pointed toward the birth canal in a "diving position" (Figure 14-14). This is the position which presents the least resistance; abnormal positions may cause calving difficulty known as dystocia (discussed thoroughly in Calving Difficulty section).

Figure 14-14. NORMAL CALVING POSITION.

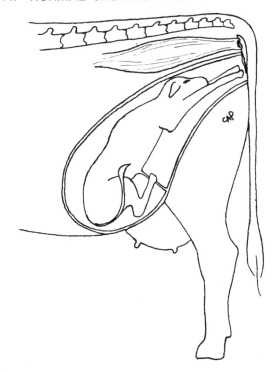

In normal birth, a fluid filled sac made of mucous membrane will appear some time after visible labor begins. The time will be extremely variable, but will normally fall within a range of 30 minutes to 2 hours. The bag will rupture which is commonly known as "breaking water". A short time afterward (15 minutes to 1 hour usually) a second fluid sac will appear. The second sac contains the fetus. It too will rupture, and then the fetus will appear . . . first the feet, and then the head. Shortly after the head appears, total expulsion should occur.

Figure 14-15. Parturition (Photos courtesy of A.T. Ralston, Oregon State Univ.).

Figure 14-15a.   Cow going into labor.

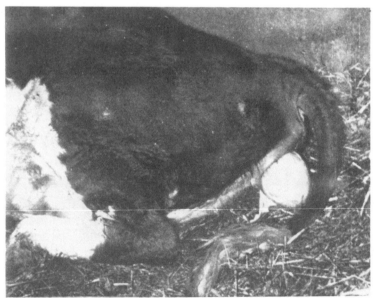

Figure 14-15b.   Cow passing the first "water bag".   It will eventually rupture, which lubricates the birth canal.   The second water bag will contain the calf.

Figure 14-15c. Calf beginning to emerge from a cow. Normal presentation is a "diving" position for the calf (see Figure 14-14.). Thus, the front feet are the first part of the calf that will be seen. Also, the feet should be pointed down . . . if they are not, or if the hind feet appear, this is unequivocal evidence that an abnormal presentation is occurring, and a veterinarian or otherwise experienced person in assisting difficult birth should be contacted without delay.

The cow will usually eat what is left of the placenta (afterbirth) and umbilical cord. She will then lick the mucus off the calf, which not only dries the calf but stimulates it to breathe. Within a period of a few minutes up to 3 or 4 hours, the calf will nurse its mother. This is extremely important as the colostrum (first milk) contains a high concentration of antibodies. It is imperative that the calf get these antibodies because cattle are unable to pass antibodies across the placenta. Thus, a newborn calf is completely vulnerable to all bovine diseases. Humans, on the other hand, can pass antibodies across the womb and thus nursing is not required.

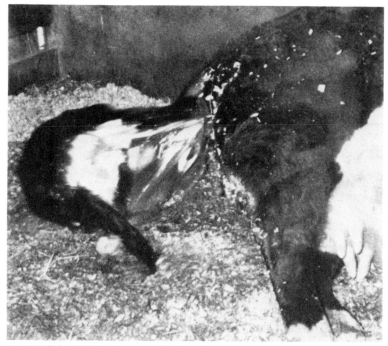

Figure 14-15d. Calf being expelled. After being expelled it is imperative that the calf begin breathing since the umbilical cord will have been ruptured, thereby cutting off the flow of oxygen. The cow will lick the calf which stimulates it to breathe (as well as dries it off). Occasionally there may be placenta or mucus lodged in the nostrils which may need to be cleared. If the calf is weak due to difficult birth etc., it may have difficulty breathing, which will call for resuscitation. This is accomplished by covering the mouth and blowing into the nostrils until the chest rises (repeat every 6 to 7 seconds). When possible, it is advisable to dip the calf's navel in iodine to reduce the possibility of infection.

DYSTOCIA OR DIFFICULT BIRTH

The practice of calving 2-year-old heifers, as well as the advent of using large European breed bulls on smaller British breed cows has intensified problems with calving. Birthweight is positively correlated with growth, and so anytime bulls are selected for superior growth characteristics (weaning weight, feedlot gain, etc.), greater birthweights can be expected. With greater birthweights will also come more difficult birth. The major problem, however, is with 2-year-old heifers.

Most texts refer to difficult birth as a matter for veterinarians. This is certainly sound advice, but under range conditions is somewhat idealistic. The following discussion is designed to help the layman render assistance to his own cattle.

GUIDELINES FOR CALVING ASSISTANCE

Thomas D. Price M.S.
Director Beef Management Systems
American Breeder's Service

(1.)   Sanitation is of major importance. All persons entering a heifer should first wash with a strong germicidal soap. Only OB chains should be used; not ropes, as ropes cannot be disinfected. OB chains should be soaked in germicide solution or at least placed in boiling water between heifers. This may be difficult to do, but must be done or uterine infections will result.

These procedures are necessary when the attendant or chains come into direct contact with the heifer. If the calf is hanging out and all that is needed is a grip on the calf, then obviously these proceedures aren't critical.

(2.)   Wash the heifer if possible. Do not use a strong soap as it will irritate the reproductive tract. Ivory soap does not irritate mucous membranes.

(3.)   Tie the heifer's tail around her neck with a piece of twine. Do not tie the tail to anything but the heifer. - That makes sense.

(4.)   To place chains on the calf. Front feet - above and below the fetlock with 2 half hitches. This gives equal pressure to the calf's leg. If applied correctly you shouldn't damage a leg. Do not chain both legs together; chain them separately. Direction of pull should be from the bottom of the leg. Pulling from the top puts a torsional stress on the leg.

If a calf is coming backwards (posterior with rear legs out the vulva), the chains should be applied as far up the leg as possible. One half hitch is enough. This will insure a faster delivery of the calf, which is very important because when the calf's umbilicus cramps against the pelvic brim he quits getting oxygen and CO_2 exchange. This will cause him to gasp for air, but all the calf will get is fluid, and consequently he will drown.

(5.)   Pulling the calf. Anterior presentation. Place the chains as shown (Figure 14-16) and proceed in pulling the calf (with calf jack or come-along). The most important thing to remember is that the pull force should be steady and downward, especially after the calf is part way out.

Figure 14-16. CORRECT PLACEMENT OF CHAINS FOR PULLING CALF.

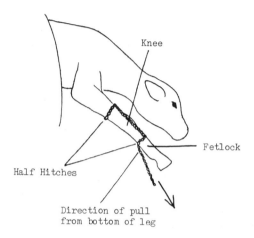

Figure 14-17. PROPER DIRECTION OF PULL DURING CALVING ASSISTANCE.

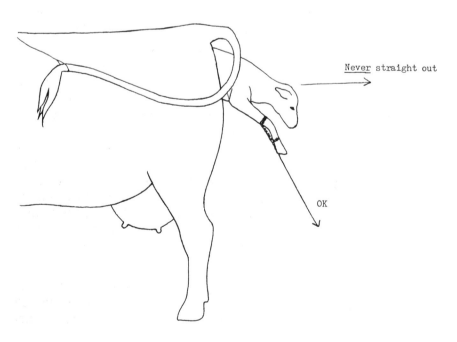

If the calf is pulled straight out a hip lock will result every time. Reason - the shape of the cow's pelvis and thighs of the calf.

The cow's pelvis narrows ventrally, but the calf is wider ventrally. If you pull down on the calf, the calf's hips go up. If you pull straight out the wider part of the calf's hips fall down into the narrow portion of the cow's pelvis . . . thus, the hip lock.

Figure 14-18. GEOMETRICAL SHAPES OF COW PELVIS AND CALF (which can create hip lock).

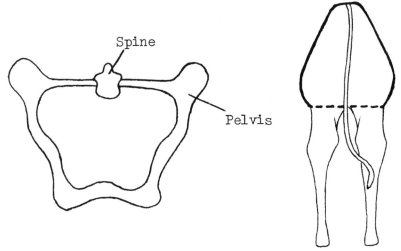

Posterior presentation. Pull the calf straight out, and don't waste time. As mentioned before, when the hind legs appear out the vulva, there is a good chance that the umbilical cord is being cramped at the brink of the pelvis.

HELPFUL HINTS

A calf can be pulled with the heifer standing or laying on one side. It is easiest to maneuver the calf with the heifer standing, but most times a heifer will go down when good pressure is applied.

How much pressure to apply? This takes a little experience Most generally, when the calf is up in the pelvic canal you should be able to get a hand in between the calf's head and the pelvis. Most inexperienced persons will call a pull harder than it really is, but common sense is usually the best guide. With experience you will know when a calf can or cannot be pulled.

If you cannot get the head and both legs, including elbows, into and over the pelvic brim it will be a hard pull. If the calf's head is compacted against the top of the pelvis, this is also a good indication of a hard pull.

A posterior pull: Most times the cervix does not dilate with this presentation. First thing is to dilate it with manual stimulation. If that doesn't work, hormones (oxytocin, etc.) will have to be used. If and when the cervix is dilated the thighs should be able to come through without pulling the cervix with them. It is easily detected if the calf will come through the pelvic canal.

Special note. A posterior calf is pulling against the hair and therefore use lots of OB lubricant. You can help the calf through the cervix and dilate the cervix simultaneously by pulling the calf and pushing the cervix over the calf's body at the same time. Again, be sanitary, but don't be afraid of working in the cow. If you can't get the cervix dilated, then a caescarian will have to be performed, which really does necessitate the services of a veterinarian.

Hip Locks. If a calf is pulled correctly (anterior presentation), you should have very few hip locks - but if one occurs, you might try this:
At this point the heifer will be down. Rotate the heifer from one side to the other; at the same time applying pressure to the calf with the pullers approaching a parallel with her rear legs. This will usually do the trick if there is no other way to alleviate the lock.

This process turns or rotates the pelvis while keeping the calf in one position. It is much easier than trying to rotate the calf.

When to assist a heifer?
Generally, if a heifer has been in first degree labor (laying down, occasionally straining), and has not shown her water bag within 2 hours, she should be examined. If a heifer has not calved within 1-2 hours after showing her water bag, again she should be examined. However, this is quite variable.

Example: A heifer that has been straining with much effort should be examined or at least watched more carefully than a heifer that hasn't given much effort. If a heifer has her calf's head out with legs, observe to see if the calf's head is swollen. There will usually be some swelling, but if the tongue, nose, etc. are quite noticeably enlarged, this is an indication that she needs help.

Always examine if you suspect a malpresentation; i.e. backwards, leg back, head back, etc. When going in to examine the heifer remember, rear legs bend in two directions; front legs only bend in one direction. Also, remember that in a posterior presentation, the calf's feet will usually be turned up.

Overview. Rendering calving assistance is basically an art learned through experience. The big thing is common sense and knowledge of which way to maximize the size of the pelvic opening and minimize the resistance of the calf.

Much has been said concerning the problems caused by overzealous assistance. Certainly an attempt to pull a calf before the cervix is dilated etc. can cause serious problems. However, assistance and examination should not be confused. If sanitary conditions are maintained, examination rarely causes any problems. If an extremely serious situation such as breech exists, early examination often makes the difference between correcting the situation and or obtaining professional help and losing the calf and dam.

PROBLEMS AFTER CALVING - RETAINED PLACENTA & PROLAPSE

Retained placenta is a problem which may or may not be associated with dystocia. The condition is very unsightly and so many stockmen are prone to want to manually remove the placenta. This may not be a wise thing to do. In a study reported by the Texas Agricultural Experiment Station (Beverly, 1975), cows that received antibiotic treatments but did not have the placenta manually removed, had a 79% conception at first rebreeding. Cows that were treated with antibiotics and had the placenta manually removed had only a 39% conception at rebreeding. Cows that received no antibiotics or assistance had a 50% rebreeding.

The problem with removing the placenta manually is that many of the caruncles are still intact. The function of the caruncle is to pass nutrients from the uterus to the placenta, and so when torn away, the pathway formerly used for nutrients, is laid open for bacterial invasion. Septicemia, or blood poisoning can result.

A fairly common procedure in the past has been to place a sulfa bolus into the uterus. However, at the present time, some reproduction experts are of the opinion that the dirt and debris that is inevitably introduced along with the bolus, creates more infection than it cures. The preferred treatment is now considered to be using an insemination pipette for the infusion of antibiotics in a prepared solution (48-72 hours after parturition). In addition, injections of antibiotics are also commonly used (Beverly, 1975).

If retained placenta becomes overly common, it may be an indication of disease. Likewise, it may also be a sign of malnutrition.

Another problem which can occur, particularly in heifers, is prolapsed uterus. The uterus becomes turned inside out and protrudes from the vulva (Figure 6-20). Prolapse can be a result of dystocia or normal calving. Also, prolapse can occur spontaneously, particularly in some breeds or lines of cattle. (In the feedlot, incidences of prolapse are apparently increased through the use of implanted hormones.)

Treatment commonly consists of pushing the uterus back into the body, and suturing the vulva. If the uterus has not been torn, and there is no contamination or infection, the animal will usually recover. The uterus may prolapse again spontaneously, and the chances of the cow being infertile are relatively high. For this reason, prolapsed animals should always be culled from the breeding herd (in addition, propensity to prolapse is often a genetic fault).

MANAGEMENT PRACTICES AS AN AID IN REDUCING DYSTOCIA

Thomas D. Price M.S.
Director Beef Management Systems
American Breeder's Service

Research has shown that the primary cause of calving difficulty results from a disproportion in size between the dam and the fetus. Calving problems arising because of abnormal presentations and postures of the calf are infrequent and not repeatable.

Birth weight of the calf is the most influential trait affecting calving difficulty. There are numerous factors that influence birth weight, and consequently calving difficulty. However, the best known method for keeping birth weights down and calving difficulties at a minimum is to use bulls known to sire small calves at birth.

Nongenetic factors, such as nutrition can influence birth weight and calving difficulty. Research reports show that heifers receiving low levels of feed prior to calving give birth to smaller calves, but calving difficulty and calf losses are not reduced. Heifers which are severely undernourished will give birth to weak calves that have a high mortality rate. Furthermore, poorly fed heifers are slow to come back into heat after calving and more difficult to get pregnant.

Heifers that receive too much feed before calving also have problems.. They become too fat, in addition to their calves being heavier, and consequently they have a high incidence of calving problems and calf deaths. It is recommended that heifers reach 80 to 85% of their mature weight by the time they are 2 years of age.

Second in importance of factors affecting calving difficulty is the size of the dam's pelvic opening. Figure 14-19 demonstrates the relationship of pelvic area measurements taken near the time of breeding and calving difficulty. Note, that as pelvic size increases the percent calving difficulty decreases.

Figure 14-19. PLOT OF PELVIC AREA & % DYSTOCIA IN ANGUS & ANGUS X HEREFORD HEIFERS.*

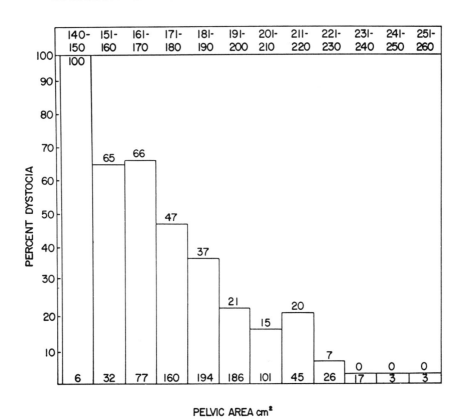

PELVIC AREA cm²

*Price, Research Highlights, Colorado State Univ. 1973.

Pelvic measurements can be used as a management tool to cull heifers that have an extremely small pelvis at or near the time of breeding. This practice will eliminate many heifers that would experience severe calving problems.

Selecting heifers alone for large pelvic size will not eliminate the occurance of calving difficulty, as heifers with large pelvic size, generally produce large calves at birth. Therefore, what is gained by increasing pelvic size is sometimes offset by the heifer producing a bigger calf. This explains why many heifers with large pelvic openings experience difficult calvings and why some heifers with small pelvic size do not. If birth weights were predictable prior to calving, which they are not, the effectiveness of selecting heifers against calving problems would be greatly enhanced.

For the commercial cow-calf operator, the decision of which breed and bull to use for calving ease must take several economic factors into consideration. For example, using a Jersey bull on virgin heifers will keep calving difficulties at a minimum, but growth rate of the progeny will be inferior and calves will be docked when sold. On the other hand selecting an Angus bull, proven superior for calving ease, will produce slightly larger calves that may experience a few more difficult births. However, the calves will have a higher growth rate and will not receive the price discrimination that the Jersey calves will encounter. Therefore, under most circumstances using an easy calving Angus bull on virgin heifers will provide the greatest economic return.

The selection of a bull for calving ease must be based on objective information. Selection of so-called "calving ease bulls" on visual appraisal has disappointed many a producer in terms of dead calves at birth.

There are large differences between bulls of any breed in their ability to sire small calves at birth that are born with less difficulty. Table 14-1 represents data collected on a group of Angus bulls

Table 14-1. BIRTH WEIGHT AND DYSTOCIA IN ANGUS BULL PROGENY.

No. Bulls Tested	12
No. Calves	511
Birth Weight	58-67 lb
% Dystocia	9-47%

*Price, 1973.

that were progeny tested to measure their calves' birth weight and calving ease ability. The sire group averages for birth weight ranged from 58 lb to 67 lb and calving difficulty from a high incidence of 47% to a low of 9%. Without a progeny test it would have been impossible to sort out which bulls were superior for calving ease.

Data from the American Simmental Association has also shown the wide variation between bulls for progeny birth weight and calving ease. Furthermore, there are also differences in bulls for their ability to produce daughters that calve with less difficulty. Data from the Simmental Sire Summary shows the average trait ratio for calving ease score to vary 40 percentage points between daughters of different sires. This means that we can select bulls to produce daughters that calve with less difficulty. Presumably, this difference between sires may be in their ability to produce daughters that have larger pelvic size and the ability to hold down birth weights.

The selection of bulls for calving ease is easy when there is access to using progeny proven bulls. But what about those cattlemen that don't have this availabilty? What tools for sire selection should they use?

The first criteria should be to select a breed known for having small birth weights. Then select a bull within the breed that had a small birth weight as compared to the birth weights of other bulls in the same group. Additional confidence can be attained if birth weight records on close relatives are also available. The heritability of birth weight is moderate, and therefore a bull selected for his small birth weight and relatives' small birth weight, will tend to produce calves with small birth weights. The selection of bulls based on their conformation and mature size can be extremely misleading and therefore it is recommended that very little attention be given to this method of evaluation.

NUTRITIONAL ASPECTS OF REPRODUCTION

More reproductive failures are due to nutrition than any other cause. There are many mineral and vitamin deficiencies which can directly or indirectly affect reproduction. However, with the possible exception of phosphorous (see page 202), most deprivations occur in the form of malnutrition. Malnutrition, of course, may result as a simple matter of not enough to eat. Probably equally as common, malnutrition may occur as a reliance on low quality roughages without adequate protein supplementation. (The reader is urged to review the section on supplementation beginning on page 29; technical discussion is provided beginning on page 195).

The following section is devoted to the two most economically important nutrients, energy and protein.* Only the female will be considered, as reproduction does not create any special nutrient requirements for the bull.

Figure 14-20. Most reproductive problems are caused by simple malnutrition. In years past, inadequate phosphorous was a major nutritional cause of poor reproduction.

ENERGY AND GESTATION

Dry, pregnant, mature cows need about 8.0 lb of TDN to maintain their own bodyweight and produce a healthy calf. That figure will vary up and down, depending upon the actual frame size of the cow, but 8.0 lb is the accepted rule of thumb for beef cows.

*Information on the effects of minerals and vitamins is covered in Chap. 12.

If a cow is deprived of energy, the effect it will have on the calf will depend upon the seriousness of the deprivation, the stage of gestation that the deprivation occurs, and the condition of the cow. If the cow is in good condition, mild deprivations of energy (up to 20% NRC) can be tolerated quite well with no significant effect on birthweight or survivability of the calf. Apparently the fetus has a priority for energy as long as the dam has energy reserves (fat). Once the energy reserves of the dam are exhausted, the fetus apparently loses its priority for energy. The condition of the cow must be quite poor, however, before any significant difference in birthweight will occur (Price, 1976; Ewing, 1959).

The effect of serious deprivations of energy (up to 50% NRC) will also depend somewhat on the condition of the cow, but the timing of the deprivation becomes the crucial factor. Most of the fetal growth takes place late in the gestation period, and indeed, this is the most crucial time as far as deprivations are concerned. Cows which are deprived of energy during this period will have calves with lighter birth weights and reduced survivability (Wiltbank, 1966).

If a serious deprivation of energy occurs early in the gestation period, ill effects can be avoided if the cattle are fed properly later on during the gestation period. In a study conducted in Wyoming, Hereford cows in moderate flesh were fed only 50% the NRC requirement for energy starting 100 days before calving.

Table 14-2. CALVING AND REPRODUCTIVE PERFORMANCE OF COWS RECEIVING SUBSTANDARD ENERGY LEVELS.*

	Energy Levels	
	low-high	low-low
Birth weight	66.8 lb	58.7 lb
% calving difficulty	9.5%	9.5%
% calves alive at birth	100%	90.5%
% calves alive at weaning	100%	71.4%
% calves having scours	33%	52.0%
% calves died due to scours	0%	19.1%
Weaning weight	320 lb	294 lb
Milk production of cows	12.1 lb/day	9.0 lb/day
Interval to 1st estrus	41.6 days	49.5 days

*Corah, 1974.

Half the cows were then fed 120% the NRC recommendations 30 days before calving, and the other half continued on the 50% ration. At calving both groups were switched to a ration conforming to the NRC lactation recommendation. Results are shown in Table 14-2.

On the basis of this data it is apparent that energy deprivations are at least partially reversible. Protein deprivations, however, do not appear to be reversible (see next section). This is important to remember since most energy deprivations occur as malnutrition; i.e. the research trials reported in this section reduced energy, but not protein, which is not the same as not getting enough to eat.

If energy intake is excessive, calf weight may increase slightly (Price, 1978). If the cow becomes obese, dystocia may result; not because of the increased calf weight, but because of fat deposited in the birth canal.

PROTEIN AND GESTATION

National Research Council recommendations are set at approximately 1.0 lb of crude protein for dry, pregnant cows. A number of trials have shown that light, weak calves with reduced survivability will result when these recommendations are not met (Wallace, 1964; Ewing, 1959; and Church, 1956). While it has been demonstrated that adverse effects of energy deprivation can be overcome if proper feeding is practiced late in the gestation period, there has been very little work with protein in this regard. From the information that is available, it appears that protein deprivations may not be totally reversible. If cows suffer protein deficiencies early in the gestation period, there is evidence to show that smaller, weaker calves may result, even if adequate protein is fed later on during gestation (Riggs and Marion, 1974).

NUTRITION AND THE ESTROUS CYCLE

The estrous cycle is the very basis of cow-calf reproduction and is greatly affected by plane of nutrition. Wiltbank (1965, 1966) has shown that it is energy that affects the estrous cycle. Protein does not directly affect estrus. Specifically, it is the secretion of gonadotropins (hormones which induce the ovarian cycle) which are related to energy level. If the cow is energy deficient, the secretion of gonadotropins (and subsequently estrus) will be delayed.

Precalving energy levels affect return to estrus only to the extent of body condition. Cows in "good" condition, will tend to breed back quicker than cows in "poor" condition, when fed identical rations. However, if cows in "good" condition are fed substandard energy levels during the rebreeding period, conception rates will be adversely affected. If cows in "poor" condition are fed an increased energy level during the breeding period* conception rates can be as good as cows in heavier flesh. Table 14-3 represents the results of a study in which cows were fed either 50% NRC (low) or 100% NRC (high) TDN levels before and after calving (Wiltbank, 1966).

Table 14-3. REBREEDING RATES OF COWS FED HIGH & LOW TDN LEVELS BEFORE & AFTER CALVING.

TDN Level		Rebreeding
Before Calving	After Calving	Conception Rate
high	high	95%
high	low	70%
low	high	95%
low	low	20%

Clearly, energy level drastically affects the estrous cycle.

NUTRITION AND LACTATION

When a beef cow calves and begins lactating, protein requirements increase by approximately 100%, and energy (and phosphorous) requirements increase by nearly 70%. In order to accommodate these radical increases, range cow herds are bred to calve in the spring or fall . . . when the forage is green (at its highest nutritional level). If due to drouth or other causes, either the quantity or quality of forage is limited, some range cows may not be able to meet their lactation requirements.

Lactation has priority over the other body functions for nutrients. Most notably, lactation has priority over the estrous cycle. When the lactation requirement for energy cannot be met, body reserves (fat) are called on to make up the deficit. (This is why heavy milking cows are often in poor flesh at weaning.) This creates an energy deprivation. As explained previously, energy deprivations adversley affect the estrous cycle. Witness of this is borne out by the rebreeding problems encountered by dairymen with high producing cows. A cow giving 100 lb of milk simply cannot eat enough feed to meet her lactation requirement. As mentioned previously, a range

---

*With sheep this practice is known as "flushing" and is done to increase the number of twin lambs, as well as overall conception.

cow giving 12-16 lb of milk, may not be able to find enough feed to meet her requirement. Indeed, in the ranching areas of the West and Southwest, it is well known that reduced calf crops are to be expected during drouth years.

Figure 14-21. Under range conditions found in much of the west and southwestern U.S., a cow with an increased ability to milk may be at a disadvantage. Such a cow may not be able to meet her lactation requirements for nutrients either because of simple availability of forage, or high fiber. When lactation requirements are not met, rebreeding conception rates are adversely affected.

The effect of range condition becomes more profound as the genetic potential for lactation increases. In the mid and late 1960's, there was attention given to crossing beef breeds with dairy breeds (Brown Swiss, in particular) as a method of increasing milk production in the cow herd (and subsequent weaning weight of the calves). Ranchers that attempted such programs ran into rebreeding problems (Riggs, 1973; personal communication*). Totusek (1975) compared Holstein, Hereford, and Holstein X Hereford heifers and cows under Oklahoma range conditions. When supplemented heavily with concentrates, the Holstein and Holstein X Hereford females had conception rates similar to the Herefords. But when supplemented at moderate levels, conception rates dropped off significantly.

*J.K. Riggs, Dept. Animal Science, Texas A&M Univ.

NUTRITIONAL REQUIREMENTS OF FIRST CALF HEIFERS

Efficient production demands that heifers calve as two-year olds. This means that heifers must be ovulating at 15 months of age. Puberty in heifers (the onset of estrus), is determined by body size as well as age (Allden, 1970; Wiltbank, 1958). For a heifer to begin cycling she must weigh from about 550-700 lb, depending upon breed and/or frame size (Wiltbank, 1974). In order to reach that weight by 15 months of age, most heifers will have to gain from 1.0 to 1.5 lb/day after weaning. After breeding, heifers need to continue to gain about 1.0 lb/day to allow for the growth of their fetus, as well as their own growth. In most ranching and some farming situations, this means that additional feed will have to be provided during at least part of this period. In providing extra feed, care should be exercised soas not to allow the heifers to gain more than the recommended amounts. In most situations, if heifers gain 1.75 lb or more/day, at least part of the gain will be fat. Fat, as mentioned previously, can increase dystocia. If heifers gain less than the recommended amounts, their calves will be smaller and weaker. The smaller calves will not reduce dystocia (Wiltbank, 1974), as the heifers will also be smaller (due to reduced growth).

After calving, severe nutritional stresses are placed on two-year old heifers. As explained in the previous section, lactation essentially doubles the nutrient requirements. A first calf heifer must not only meet the increased lactation requirements, but must also meet her continued growth requirement. Growth and lactation both have priority over estrus for nutrients (energy). Therefore, if first calf heifers are to rebreed, they must be given access to the best pasture available and/or given extra feed. As a practical matter, lactating heifers cannot be overfed at this time. To reduce the amount of feed required and otherwise increase rebreeding conception rates, some operations practice early weaning (see page 28). By removing the nutritional demands of lactation, heifers are more able to meet their growth requirements (under grazing conditions), and therefore return to estrus will be enhanced. In addition, Wiltbank (1958) has shown that the stimulus of suckling also inhibits the ovarian cycle.

CONCLUSIONS

1. Energy has a profound effect upon estrus. If cows or heifers are energy deficient during the breeding period, conception rate will be substantially reduced.

2. Cows may undergo energy deprivation early in gestation without adverse affect . . . if additional energy is provided during late gestation.

3. Protein deprivations may not be totally reversible.

CHAPTER 15   BEEF CATTLE BREEDS

To feedlot operators the selection of breeds is a relatively simple topic. They know how particular breeds or crosses of breeds perform, and they know approximately what the carcass will be like. Their decision on breeds is therefore a trade-off of how the cattle will perform versus what the local packer buyers will pay for the finished animal.

To the cow-calf operator, beef cattle breeds is a very controversial topic. This is primarily due to the fact that cow-calf operators are not able to measure performance of their cattle as accurately as feedlot operators . . . there are too many variables beyond their control. In addition, some of the traits they look for, such as mothering ability, etc., do not lend themselves to objective evaluation. The selection of a breed or a particular cross therefore becomes a somewhat perplexing decision. In almost all situations, the final decision is (or should be) a matter of compromise, weighing all the environmental and marketing conditions together (see also page 23).

BREED DESCRIPTIONS

The following is a series of brief descriptions of the more popular beef breeds. The author has made every effort to be completely objective in evaluating the advantages and weaknesses of the breeds. As mentioned earlier, however, beef breeds is a controversial topic, and objective descriptions are certain to offend individual purebred breeders and/or breed associations. The reader should therefore remember that <u>all breeds have weaknesses</u> (as well as advantages), and that it is the author's duty to present all the information concerning each breed.

THE CLASSIC "U.S." BRITISH BREEDS

The classic British breeds, Hereford, Angus, and Shorthorn were brought to the U.S. in the 1800's. For nearly a century these 3 breeds formed the basis of nearly all the beef cattle raised in the U.S.

HEREFORD (Polled and Horned)
The red bodied, white faced Hereford is the most numerous of all beef breeds in the U.S., and at one time was undoubtedly the most popular breed. At the present time, there are wide variations of opinion concerning the value of Hereford cattle.

The differences of opinion concerning Hereford cattle (in the author's opinion), is due to a wide genetic variation within the breed. Polled and horned varieties are available . . . indeed, they are considered by many to be two different breeds. There are far greater differences between Hereford cattle, however, than whether they have horns or not. There are wide variations for growth, hardiness, and resistance to disease.

Figure 15-1. Grade Hereford steer. Follett Feeders, Follett, TX.

These differences are attributable to the breeding management of individual herds. Some breeders have practiced sound principles of outbreeding, while others have not. Unfortunately, there are herds within the U.S. that are highly inbred. Inbreeding decreases the hardiness and ability to resist disease (as well as the growth rate), which has created ill feelings for the breed by stocker operators and feedlots receiving cattle from these herds. While a detriment to the breed in general, this has been an asset to the owners of well bred herds, as it has made their animals much more valuable.

Herefords are relatively well adapted to most of the grazing lands of the western U.S. (particularly the mountainous areas), and it has been this adaptablility that has apparently made the breed so popular. The Hereford breed is quite fertile, in particular, Hereford cow herds are able to remain relatively fertile under typically sparse forage conditions of western rangelands. This is probably due to the fact that Hereford cows yield less milk than most other breeds (Melton, 1967), which means they can retain more energy for ovarian cycling (page 277 & 278).

Hereford cows generally make excellent mothers . . . another factor extremely important under range conditions. They are usually quite .conscious of the whereabouts of their calves, and are normally very quick to defend them.

ANGUS

The Angus breed was brought to the U.S. in the 1880's, and has been extremely popular ever since. The physical features are the dominant black color the breed is famous for (there is also a recessive red color), a smaller mature size than most other breeds, and an absence of horns (Angus are naturally polled).

A very profound attribute of Angus cattle is the breed's propensity to deposit intramuscular fat (to marble). USDA quality grades are based primarily on marbling, and the ability of Angus to marble has had a significant impact upon their value as feedlot animals.

Figure 15-2. Grade Angus steer. Follett Feeders, Follett, TX.

The small size has enabled the Angus to fit into a number of special situations. Mature size is related to calf birthweight and thus the Angus are small as calves. For that reason, Angus bulls are often put on first calf heifers of other breeds so as to reduce calving problems. Also, many beef cattle authorities believe that in certain instances it is better to use smaller cows since they have a lower feed requirement. When smaller cows are desired, it is most often Angus or Angus cross cows that are used.

SHORTHORN

The Shorthorn was the first of the British breeds to be brought to the U.S. Brought over in 1783, it never developed to be as popular as the Hereford or Angus. The two primary colors are red and white, with a mixture of red and white known as roan occurring about 50% of the time. The cattle, as the name would imply, have horns, and are similar in size to Herefords.

Figure 15-3. Grade Shorthorn steer (steer has been dehorned). Hansford Feedyard, Spearman, Texas.

ZEBU BREEDS

The Zebu breeds comprise a number of cattle that originated from India. Zebu cattle are actually a different genera of cattle than other breeds, or groups of breeds. Zebu being what is known as Bos indicus and the remaining commonly used breeds being Bos taurus.

Indeed, the Zebu breeds are quite different than other breeds (Figure 15-4). Characteristically they have a hump on the top of the base of the neck, exceedingly large ears, a loose fitting skin, and a prominant dewlap. Physiologically they are apparently different as they have the ability to withstand heat and tropical

Figure 15-4. Extremely high quality Brahma bull held for semen production used in artificial insemination.   Exaggerated features in
   this high quality bull exemplify the difference between Bos indicus
   and Bos taurus cattle breeds.   Courtesy American Breeder's Service.

climates much better than other breeds.   On the other hand, they
are more susceptible to cold stress than Bos taurus breeds.   Emotionally they are much more excitable than other breeds, and
are apparently more intelligent.   In the feedlot, they are more
prone to acidosis than other breeds.

THE BRAHMAN BREED
   In the U.S. there is only one recognized Zebu breed, the Brahman.   Apparently the Brahman is the result of the interbreeding
between several different Zebu breeds brought to the U.S. about
the turn of the century.
   Originally the Zebu breeds/Brahman was brought to the U.S.
for use in the hot humid areas of the Southeast, and the hot, arid
areas of the Southwest.   While very useful in that regard, the
Brahman has found more use as a crossbred animal than as a
purebred.   As explained on page 240, crossbred cattle are usually
more productive than straightbred cattle, but with Brahman
cattle the increase is much greater.   Understandably, the first
large scale crossbreeding was done with Brahman cattle, and
most of the crossbred cattle that are recognized as an individual
breed, contain Brahman blood.

Figure 15-5. Brahma cross cow and calf. Courtesy American Breeder's
Service.

An obvious reason behind crossbreeding Brahman cattle was
to improve their musculature or meat animal characteristics.
Probably more important, Brahman cattle are notoriously poor
breeders and crossbreeding results in greatly improved fertility.
Straightbred Brahman cattle can be expected to produce no more
than a 65-70% calf crop, but when crossed with most other beef
breeds, 90% calf crops can be expected. This discovery was
a milestone in the emergence of the Southeast as a beef producing
area.

CONTINENTAL (European) BREEDS

In the early 1970's a number of breeds from the European
continent were imported and heavily promoted in North America.
Restrictive importation laws (to guard against hoof and mouth
disease which is endemic in Europe) made the importation of
live animals practically impossible. To the average producer
the purchase of semen for artificial insemination was the only
real access to these breeds. This meant that the only feasible
way to obtain "purebred" animals was to repeatedly breed back
to offspring heifers from previous matings. That is, a ½ blood
heifer bred back to that same breed would have ¾ blood offspring.
Taking the ¾ blood offspring heifer and breeding back to the
same breed would produce a ⅞ offspring, etc. Speculation (in

response to all the promotion for these breeds) made ½, ¾, etc. breeding animals extremely valuable. For a while then, commercial ranchers who were quick to adopt these breeds experienced substantial monetary gains due to the intense speculation that occurred at that time. Beginning in the mid 1970's, however, the industry began evaluating these breeds much more objectively, and prices fell much more in line with traditional breeds.

To objectively evaluate these breeds one would have to conclude that in general they are genetically very useful. To fully understand how these breeds should be used, one must realize that some of these breeds are actually dual purpose cattle (used in Europe for dairy as well as beef purposes). This fact affects their potential use as range cattle.

A major part of the intensive promotion of some of these breeds centered around the fact that they can wean 600 lb calves rather consistently. Since this was about 50% greater than what was considered normal for the existing breeds at that time (400 lb), a significant amount of excitement was created.

To be sure, most of these breeds do have good growth rates; better than the average "conventional" British breeds*. This is probably due to better genetic selection on the part of European breeders. Up until relatively recently, North American cattle were unfortunately selected for a number of traits other than growth. Many of these traits were purely cosmetic; coat color, markings, etc. Probably the greatest damage came in the 1930's when U.S. breeders deemed "short, blocky, compressed" cattle to be the ideal type. In selecting for the compressed type, they actually selected for reduced growth. (A few breeders did not select for the compressed type, which is probably responsible for much of the genetic variability in North American "British" breeds.) European cattle, however, have apparently been selected primarily for productive traits. Growth rates of the European breeds would certainly contribute to the heavier weaning weights they are capable of.

The reader should remember, however, that many of these breeds are also dual purpose animals which are capable of giving up to 25 lb of milk per day. Conventional beef breeds average approximately 12-16 lb of milk, and so with about 50% extra milk, calves from dual purpose breed cows would have more than just a genetic advantage.

---

*One should realize, however, that within breeds there are tremendous differences, particularly among North American "British" breeds.

Figure 15-6. Simmental cow. Note the large udder; exemplifying the dual purpose capabilities. Courtesy American Simmental Association.

Indeed, the use of the dual purpose breeds has often been recommended as a method of increasing the milk production of existing beef herds, and certainly it would be effective. It should be remembered, however, that increasing milk production is not always an advantage in beef cows. (Please read "Nutrition and Lactation" page 278.)

As mentioned earlier, nearly all the European breeds have excellent growth rates and so they are nearly always quite useful as terminal sires; i.e. for the production of commercial calves for eventual slaughter. European crossbred cattle typically perform quite well in the feedlot, the only drawback being that they usually do not marble as well at the usual 1050-1200 lb slaughter weights. As consumer preference for heavily marbled meat changes (as it has on the West Coast), this should not be a disadvantage.

One should be cognizant that birthweight is positively correlated with growth rate. Cattle that have good growth rates, also have large calves. When using superior bulls (of any breed), the cow-calf producer can expect more difficult births due to heavier birthweights.

Some of these European breeds have been given the reputation of increasing dystocia. This is an unjust criticism as increased dystocia usually just means the cattle have a greater growth potential. Unjust or not, however, the cow-calf operator should be aware of this, and should not use these larger breeds on heifers or cows known to have calving problems.

In summary, it can be said that the European breeds are very useful. They typically make excellent terminal sires, and can contribute much to crossbreeding programs.

EUROPEAN DUAL PURPOSE BREEDS

SIMMENTAL

The Swiss Simmental has undoubtedly been the most popular of the Continental breeds. The breed has excellent growth characteristics, and has been used extensively both as a terminal sire, and to "improve" existing breeding herds. Outside of the genetic merits Simmental has, undoubtedly another reason for the breed's popularity has been the fact that it's a very beautiful breed. Probably more importantly, when crossbred with Hereford cattle, the offspring look very much like purebred Herefords . . . the main difference being a lighter, or more golden color, with occasional white markings on the body. Indeed, the traditional Hereford cattlemen find the purebred Simmental very appealing. A large, horned breed with a white face, Simmental coat color varies from a golden straw to a light reddish orange, with intermittent white markings.

Figure 15-7. Simmental X Hereford cow with ¾ blood Simmental calf. The similarity to Hereford color patterns has undoubtedly been part of the popularity of the Simmental breed. Courtesy American Breeder's Service.

MAINE ANJOU

The Maine Anjou breed originated in western France. It is a large breed, usually a dark reddish brown with a white belly and occasional white markings on the body. The breed has primarily been used as a terminal sire in North America.

Figure 15-8a&b. Maine Anjou cow and heifer calf (above). Maine Anjou bull (below). Courtesy American Breeder's Service.

TARANTAISE

The French Tarantaise is much smaller than the other European breeds, being very similar to the Jersey breed in size, coloration, and conformation of the head. The breed is apparently very fertile, and has found its main North American use in crossbred breeding herds. Being small in size, but good milkers, the cows are said to produce calves that are a higher percentage of the cow's body-weight than most beef breed cows.

Figure 15-9. Half blood Tarantaise cow (½ Tarantaise X ¼ Hereford X ¼ Angus). Note the Jersey like head, the large, full udder, and apparent small size. Courtesy American Breeder's Service.

EUROPEAN BEEF BREEDS

CHAROLOIS

The importation of the white or cream colored French Charolois breed actually preceded what is generally considered the "exotic" boom of the 1970's. A large, muscular, polled breed, the Charolois has extremely good growth rates, and was quickly accepted by U.S. and Canadian cattlemen. It has been very successful as a terminal sire, and has contributed much to crossbreeding programs. Apparently crossbred Charolois females make good range cows . . . producing large, growthy calves.

Figure 15-10. Half blood Charolois calf out of Hereford X Angus cow. Courtesy American Breeder's Service.

LIMOUSINE

The Limousine is a horned, medium size breed also originating from France. The color varies from golden straw to a reddish brown, the latter color being somewhat more common. One of the most distinctive features is that the body appears to be somewhat longer than other beef breeds.

The breed is best known for its carcass traits. A very muscular breed, the Limousine carries relatively little external fat and therefore produces very trim meaty carcasses.

Figure 15-11a. Limousine X Hereford steers (Limousine used as a terminal cross). Courtesy American Breeder's Service.

Figure 15-11b. Purebred Limousine bull. Courtesy American Breeder's Service.

CHIANIA

While the Chiania has been listed under the beef breeds, it was originally used for draft purposes. Italian in origin, their use as draft animals dates all the way back to the Roman Empire. Considering the enormous size of these animals (Chiania are

Figure 15-12. Chiania yearling steer in the feedlot. Note the long legs and height compared to the Angus. Lowe Feedlot, Holly, Colorado.

the largest cattle in the world), and the graceful, almost thoroughbred appearance of the legs, the Chiania is certainly well suited for draft purposes. Indeed, when handled constantly, Chiania develop a very gentle disposition which would also be a great attribute for draft purposes.

The Chiania breed has been very popular in Latin America. Because of its white and other apparent Zebu like features, it was theorized that the breed would do well under tropical conditions. The only drawback to the cattle at this point is that due to the length of the legs and their swiftness, they can be very difficult to herd when the animals are not used to being handled; i.e. oftimes they can outrun a horse and rider.

RECOGNIZED BREEDS DERIVED BY CROSSBREEDING

There are a number of breeds that have been developed in recent times by crossing existing breeds. In most cases these have been matings between Brahman and British or European breeds. Used primarily in the South and Southwest, the breeds have been very successful.

When crossed with Brahma, all the resultant breeds retain most of the Brahman physical characteristics; e.g. large floppy ears, dewlaps, characteristic horns, etc. In most cases the new breed carries the coat color of the British or European breed. Typically the crossbreed carries the disposition of the Brahman breed; i.e. when handled gently they can become very docile, but when handled roughly become nervous and excitable.

SANTA GERTRUDIS

The Santa Gertrudis breed, developed by the famed King Ranch, was the first crossbreed to gain wide acceptance in the U.S. as a recognized breed. Originally the two parent breeds (Shorthorn and Brahma) were present on the ranch as purebreds. The development of the crossbreed came out of necessity as neither breed was particularly productive by itself. The Brahman breed has its inherent fertility problem, and the Shorthorn simply was not adapted to the intense heat and humidity of the South Texas Gulf Coast. By crossing the two breeds, both fertility and rate of gain were increased dramatically. The cross ratio of ⅝ Shorthorn, and ⅜ Brahma was set as the standard for the breed. The

Figure 15-13. Grade Santa Gertrudis steer.

resulting animal carries most of the Brahman characteristics except for the red coat color, and an increased size and muscularity.

BRANGUS

Next to the Santa Gertrudis, the Brangus breed is probably the most popular of the Brahman crossbreeds. The breed is standardized at ⅜ Brahma, ⅝ Angus ratio, carries the black coat color of Angus, and is somewhat smaller than the other Brahma crosses. Due to the high proportion of Angus genes, Brangus cattle marble and finish quicker than other Brahman crosses in the feedlot.

BRAFORD

Braford, a cross between Hereford and Brahma cattle, is a extremely popular cross, particularly in South and East Texas. There is a Braford Breed Registry, and the official percentages are ⅝ Hereford and ⅜ Brahma. In commercial cattle circles, however, most any kind of cross between Hereford and Brahma is usually called a Braford. First cross F1 heifers (½ Hereford - ½ Brahma) are very much in demand as herd replacements in the southern parts of Texas.

The breed usually carries the red coat color and white face of the Hereford. At times, Braford cattle will display a brindle color.

BEEFMASTER

The Beefmaster is said to be ½ Brahma, ¼ Hereford, and ¼ Shorthorn. The cattle have horns and the color is quite variable.

Figure 15-14. Beefmaster bull. Courtesy American Beefmaster Breeder's Association.

Figure 15-15. Brahmantal cow photographed in Florida. Courtesy American Breeder's Service.

BRAHMENTAL

Very recently a breed association for Brahma-Simmental cross cattle has developed. The ratio has been standardized at ⅝ Simmental - ⅜ Brahma. As might be expected, the breed typically carries the blond color pattern of the Simmental.

BARZONA

The Barzona breed was developed for use in southern Arizona, probably the hottest, most arid area in the U.S. that will support range cattle. Most breeds developed for heat tolerance utilize Brahman blood for this purpose. The Barzona breed is somewhat different as Africander, another <u>Bos indicus</u> breed (from South Africa) was used as the main genetic source of heat tolerance. Brahman blood was included, however, as Santa Gertrudis breeding was also used. British breeds included (in addition to the Shorthorn in the Santa Gertrudis) were Hereford and Angus. The breed is red in color, horned, and moderate in muscling.

Figure 15-16. Barzona cow and calf. Courtesy Barzona Breeders Association of America.

MISCELLANEOUS BREEDS SOMETIMES USED IN BEEF PRODUCTION

HOLSTEIN

The Holstein, although primarily a dairy animal, makes an important contribution to beef production. Holstein heifer calves, of course, are usually kept for milk production. The bull calves, are typically finished as feedlot animals. There are some that are sold for veal, but the number has been greatly reduced.

Indeed, Holstein steers make excellent feedlot animals. They have rates of gain that will equal or exceed any beef breed. The only drawbacks being that they cannot be fed to the choice grade economically, and that (typical of dairy breeds) they have a greater amount of cod fat. Usually Holsteins will have a carcass yield about 2 to 3 percentage points below what fed beef breeds yield; i.e. most beef breeds fed to 1050-1100 lb will dress from about 61 to 63%, whereas Holsteins fed to about 1100 lb will dress from about 58 to 60%. The net result is that packers will not pay as much for Holsteins as they will beef breeds. The trade-off, how-ever, is that their cost as calves and/or yearlings is usually lower than beef breeds.

Figure 15-17. Holstein steers being fed for slaughter. Figure 1 Feedlot, Booker, Texas.

As the demand for "choice" (well marbled) meat declines and the demand for lean meat and hamburger increases, so will the value of Holsteins. This trend really began in the mid 1970's with the emergence of Keystone Foods Corp. as a major supplier of hamburger for the McDonald's chain. At the time of writing, Keystone had in excess of 30,000 Holsteins on feed year round for processing as hamburger (the loin and rib are pulled and sold as conventional cuts).

A discussion of Holsteins would not be complete without men-tioning that there is a strain of the breed that has been developed primarily for beef purposes. Known as the Beef Fresian, the breed has never been very popular in North America. The primary reason being that it has black and white spots and therefore looks like a dairy animal. A legitimate drawback to the breed, is the great milk production. A cow or heifer with the ability to give as much milk as a Fresian simply cannot get enough to eat under most pasture conditions. Totusek (1975) tried using Hereford X Holstein cows under range conditions, and found the increased milk production to be a real detriment. As long as the cows were fed extra concentrates they weaned much larger calves than the straightbred Herefords. But when managed and fed similar to other range cattle, calf crop percentage declined drastically.

In a terminal cross situation the Beef Fresian has potential. The main drawback again being that the cattle will have black and white spots and packer buyers will ultimately associate them with Holsteins. That is, offer no more than Holstein price which is usually $3-5/cwt below the market price for conventional beef breeds.

BROWN SWISS

The Brown Swiss, like the Holstein, is a large dairy breed capable of heavy milk production. During the early 1970's a number of cattlemen tried breeding Brown Swiss into their cow herds as a means of increasing milk production. The Brown Swiss was chosen primarily because it didn't have spots and therefore looked more like a beef breed. The cattlemen that attempted to run Brown Swiss and half-blood Brown Swiss under range conditions ran into the same problems as Totusek (1975) encountered with Holsteins (see page 279).

In the feedlot, Brown Swiss steers are very similar to Holsteins; i.e. excellent growth rates and somewhat reduced carcass yields. As terminal crosses they are or could be very useful for the production of lean meat.

JERSEY

If the Jersey has a place in beef production, it is in a crossbreeding program for the production of small framed breeding cows. In the feedlot the Jersey breed has nothing to offer. Growth rates are the lowest of any breed of cattle, and due to a lack of musculature, dressing percentages are very low. In addition, the Jersey breed does not convert carotene to vitamin A for fat storage very well, and thus the fat is usually a yellow color, which most consumers find objectionable.

The one attribute the Jersey has is the ability to calve easily. For reasons that are not completely clear, dystocia is very seldom

Figure 15-18. Jersey steer in the feedlot. Notice the lack of musculature and large paunch.

a problem with Jersey cows. The Jersey is a very small breed, and when crossed with larger beef breeds, the crossbred cow is usually intermediate in size . . . but the crossbred cow is usually able to give birth to as heavy a calf as the larger parent beef breed cow. Jerseys are not such heavy milkers as other dairy breeds, and so milk production in the crossbred cow is usually not much more than the larger beef breed cow. The Jersey cross cow is what is known in the academic world as a "small, metabolically efficient" cow. In other words, she produces as much milk as a larger cow, but doesn't need as much feed due to her small size (reduced maintenance requirement). Being an easy calver, she can be bred to large terminal sires. In some situations these attributes make the use of Jersey quite desirable. The following is the result of test work done with Angus X Jersey cows bred to Charolois bulls versus straightbred Hereford and Charolois cows under confined conditions (Thomas, 1971).

Age of Cows	Breed of Sire	Dam	Weight of Cows at Calving	Calving Assist % Cows	Ave. Milk Yield	Weaning Weight	lb of Feed per lb of Calf	Calving %
3 & 4	C	A x J	697 lb	4.4%	14.7 lb	510 lb	14.7	90
5+	H	H	1109 lb	0	12.3 lb	468 lb	16.4	81
5+	C	C	1258 lb	0	14.6 lb	547 lb	15.6	65

Table 15-1. PERFORMANCE OF ANGUS X JERSEY COWS COMPARED TO HEREFORD AND CHAROLOIS.

CORRIENTE

The Corriente breed is a true breed which evolved out of the desolate areas of Mexico. The word corriente is a Spanish idiomatic expression which is best translated as cur or mongrel. Indeed, the Corriente breed is a mixed up mongrelization of cattle breeds that undoubtedly had the Texas Longhorn as its basic ancestor. Eventually, the breed emerged with the genes "fixed" and therefore the Corriente is a true breed.

Figure 15-19. Mexican Corriente steers being fed for slaughter. J&B Cattle Co., Gruver, TX.

The cattle are horned . . . indeed the size and shape of the horns are one of the breed's most distinctive characteristics. The overall size of the cattle is small, with very moderate muscling, and varied color patterns. The breed is noted for resistance to disease and being able to survive under extremely harsh conditions. Due to the breed's relatively poor gaining ability as stocker and feeder cattle, buyers used to discount the breed heavily. Due to an upsurgence in the popularity of steer roping, however, Corriente cattle currently bring a premium over growthier type cattle. In the Southwest fairly large numbers of Corriente are fed out (usually after they become too large for rodeo purposes).

TEXAS LONGHORN

The Texas Longhorn, of course, was the first breed of cattle to ever be raised in North America. Originating from the cattle

Figure 15-20. Texas Longhorn bulls being used to breed first calf heifers.

brought over by the Spaniards in the 15th century, Longhorns were run on the western plains of the U.S., and then driven north and east to the population centers for marketing. As range conditions became less harsh, the Longhorns were eventually replaced by the British breeds.

At the time of writing, the Longhorn was receiving somewhat of a resurgence in popularity. The commercial use most often seen is the use of Longhorn bulls on first calf heifers as Longhorns are reputed to be easy to calve. Most use of Longhorns, however, is more for the sake of novelty than anything else.

Figure 15-21. Crossbred Longhorn steer in the feedlot. Hansford Feedyard, Spearman, Texas.

APPENDICES

Appendix Table I
Average Nutrient Compositions (Dry Matter Basis)***

	Mois-ture	TDN[1]	NE_m[2]	NE_g[2]	Crude Prot.	Dig. Prot.	Crude Fat	Crude Fiber	Ca	Phos.
Roughages										
Alfalfa Hay 21% Fiber	15%	58%	71	40	22.0%	16.1%	3.2%	23.1%	1.89%	0.29%
Alfalfa Hay 24% Fiber	12	55	57	27	20.7	14.4	2.9	26.3	1.63	0.27
Alfalfa Hay 28% Fiber	10	53	51	18	17.1	12.4	1.7	31.7	1.58	0.27
Alfalfa Hay 34% Fiber	10	50	47	9	13.7	9.1	1.6	39.9	1.19	0.21
Alfalfa DeHy 20% Prot.	10	64	62	36	22.4	16.3	3.1	22.2	1.93	0.28
Alfalfa DeHy 17% Prot.	10	62	59	32	19.6	13.7	2.7	27.3	1.70	0.29
Barley Straw	8	41	45	7	4.1	0.8	1.8	30.8	0.37	0.11
Bremudagrass Hay, common	10	47	48	13	7.9	4.0	2.0	28.8	0.41	0.21
Bremudagrass Hay, Coastal, N. fert.	10	49	54	24	10.2	7.0	2.3	30.9	0.30	0.20
Corn Cobs, ground	10	47	48	14	2.6	--	0.4	35.7	0.12	0.04
Cottonseed Hulls	10	41	47	10	4.3	--	1.0	50.0	0.14	0.07
Milo Stover	10	57	41	13	3.6	1.2	1.2	32.3	0.64	0.12
Oat Hay	10	61	52	18	9.9	5.6	3.0	31.2	0.23	0.21
Prairie Hay	10	49	52	18	6.7	2.2	3.3	32.2	0.37	0.13
Peanut Hulls[3]	8	32	2	1	1.9	1.3	1.8	67.1	0.28	0.07
Rice Hulls	9	24	20	10	1.7	--	1.1	50.0	0.11	0.08
Wheat Straw	9	48	16	7	4.3	0.3	1.7	41.1	0.16	0.08
Silages										
Alfalfa Silage	50-75	54	60	30	17.6	12.0	3.9	31.7	1.56	0.28
Corn Silage[4]	62-72	60-70	73	43	9.1	5.6	4.8	25.1	0.38	0.33
Sorghum Silage[4]	60-78	59	67	30	7.9	4.2	4.8	26.9	0.39	0.23
Concentrates										
Barley 46-48 lb	11	83	97	64	11.9	8.9	2.1	6.0	0.07	0.44
Barley, light weight	10	70	80	53	13.4	10.2	2.3	8.2	0.07	0.36
Beet pulp, molasses dried	8	74	92	61	9.9	6.6	0.6	16.9	0.63	0.08
Corn dent, No. 2	15	91	102	61	10.7	8.2	4.5	2.4	0.22	0.31
Corn Cob Meal	10	47	90	60	8.9	6.5	3.7	10.0	0.44	0.25
Corn Gluten Feed 25% protein	9	84	83	55	27.5	24.0	2.8	8.0	0.45	0.89
Cottonseed Whole	8	91	91	54	25.7	19.0	25.4	18.8	0.15	0.75
Cottonseed Meal, 41% solvent process	8.5	75	71	45	46	41	2.3	12.2	0.17	1.22
Fat (beef tallow)	.5	--	225	141	--	--	105.5	--	--	--
Linseed Meal, solv.	9	76	81	54	40.7	34.1	1.1	10.3	0.43	0.95
(1) Milo Grain, 8% prot.	10	75	80	53	8.8	6.1	3.1	2.7	0.04	0.33
(2)	10	78	90	60	8.8	6.1	3.1	2.7	0.04	0.33
(3)	10*	85	97	64	8.8	6.3	3.1	2.7	0.04	0.33
(1) Milo Grain, 9% prot.	10	75	80	53	10.0	6.8	3.1	2.7	0.04	0.35
(2)	10	78	90	60	10.0	6.8	3.1	2.7	0.04	0.35
(3)	10*	85	97	64	10.0	6.8	3.1	2.7	0.04	0.35
(1) Milo Grain, 10% prot.	10	75	80	53	11.1	7.7	3.1	2.7	0.04	0.37
(2)	10	78	90	60	11.1	7.7	3.1	2.7	0.04	0.37
(3)	10*	85	97	64	11.1	8.1	3.1	2.7	0.04	0.37

Appendix Table I (continued)
Average Nutrient Composition (Dry Matter Basis)***

	Moisture	TDN	NE_m	NE_g	Crude Prot.	Dig. Prot.	Crude Fat	Crude Fiber	Ca	Phos.
Concentrates										
Molasses, Beet 79.5 degrees brix	32.5	70	83	53	10.5	5.5	--	--	0.66	0.22
Molasses, Cane 79.5 degrees brix	33	77	83	53	3.8	--	--	--	0.83	0.10
Oats	10	75	83	57	13.3	10.4	5.1	12.2	0.10	0.37
Soybean Oil Meal, solv. 50%	11	81	87	58	56.0	51.5	1.1	6.4	0.28	0.70
Soybean Seed, ground	10	94	108	70	42.1	37.4	20.0	5.5	0.27	0.65
Wheat Grain, 11% prot.	11	88	100	65	12.3	10.2	2.1	2.3	0.05	0.32
Wheat Grain, 12% prot.	11	88	100	65	13.7	11.4	2.0	3.1	0.05	0.43
Wheat Grain, 13% prot.	11	88	100	65	15.0	12.5	2.0	3.1	0.05	0.47
Wheat Mill Run	10	74**	87**	49**	19.4	16.1	5.8	8.4	0.09	0.91
Minerals										
Ammonium Polyphosphate	--	--	--	--	93.8	--	--	--	--	22.35
Bone Meal, steamed	--	--	--	--	5.5	4.2	0.4	1.4	31.30	14.40
Diammonium Phosphate	--	--	--	--	128.0	--	--	--	--	23.50
Dicalcium Phosphate	--	--	--	--	--	--	--	--	27.00	19.10
Disodium Phosphate	--	--	--	--	--	--	--	--	--	21.80
Ground Limestone	--	--	--	--	--	--	--	--	38.00	--
Monosodium Phosphate	--	--	--	--	--	--	--	--	--	25.80
Oyster Shell Flour	--	--	--	--	--	--	--	--	38.00	--
Rock Phosphate, D.F.	--	--	--	--	--	--	--	--	24.00	18.50
Sodium Tripolyphosphate	--	--	--	--	--	--	--	--	--	25.00
Salt	--	--	--	--	--	--	--	--	--	--
Trace Mineral Mix	--	--	--	--	--	--	--	--	--	--
Tricalcium Phosphate	--	--	--	--	--	--	--	--	33.00	18.00
Urea, 45% N	--	--	--	--	281.0	--	--	--	--	--
Ammonium Chloride	--	--	--	--	163.0	--	--	--	--	--

All minerals expressed on normal moisture basis.

[1]TDN typically overestimates the value of roughages

[2]Megacalories per hundred pounds

[3]Analysis can vary greatly depending upon the amount of immature nut meats present.

[4]Analysis can vary greatly with maturity and the amount of grain present.

(1) = ground dry   (2) = fine ground or rolled   (3) = steam flaked, popped, micronized, or high moisture harvest.

*Moisture after processing will depend upon the type of processing.

**Quite variable - depending upon the amount of dirt present.

***Adapted from Kansas Feedlot Handbook, 1970.

Appendix Table II

Daily TDN, Protein, Ca, P and Vitamin A Requirements for Growing-Finishing
Steer Calves and Yearlings.* (Dry Matter Basis)

Weight (lb)**	Daily Gain (lb)	Total Protein (lb)	Digestible Protein (lb)	TDN (lb)	Ca (g)	P (g)	Vit. A (thous. IU)
220	0	0.40	0.22	2.6	4	4	5
	1.1	0.79	0.53	4.0	14	11	6
	1.5	0.88	1.62	4.4	19	13	6
	2.0	1.01	0.73	4.6	24	16	7
	2.4	1.08	0.79	5.1	28	19	7
331	0	0.51	0.29	3.5	5	5	6
	1.1	0.97	0.62	5.5	14	12	9
	1.5	1.08	0.73	6.0	18	14	9
	2.0	1.19	0.81	6.6	23	17	9
	2.4	1.28	0.90	6.8	28	20	9
441	0	0.66	0.37	4.2	6	6	8
	1.1	1.25	0.77	7.5	14	13	12
	1.5	1.34	0.86	7.9	18	16	13
	2.0	1.34	0.88	8.2	23	18	13
	2.4	1.39	0.95	8.6	27	20	13
551	0	0.77	0.44	5.1	8	8	9
	1.5	1.36	0.86	8.8	18	16	14
	2.0	1.52	0.97	9.9	22	19	14
	2.4	1.61	1.06	10.4	26	21	14
	2.9	1.67	1.12	11.5	30	23	14
661	0	.088	0.51	5.7	9	9	10
	2.0	1.78	1.10	11.9	22	19	16
	2.4	1.80	1.14	12.3	15	22	16
	2.9	1.83	1.18	13.2	29	23	16
	3.1	1.91	1.25	13.7	31	25	16
772	0	1.01	0.57	6.4	10	10	12
	2.0	1.76	1.08	12.8	20	18	18
	2.4	1.83	1.14	13.7	23	20	18
	2.9	1.91	1.21	15.0	26	22	18
	3.1	1.98	1.25	15.4	28	24	18
882	0	1.12	0.64	7.3	11	11	13
	2.2	1.91	1.19	15.0	21	20	19
	2.6	1.91	1.19	15.4	23	21	19
	2.9	1.98	1.23	16.1	25	22	19
	3.1	2.07	1.30	17.0	26	23	19

Appendix II (continued)

Daily TDN, Protein, Ca, P and Vitamin A Requirements for Growing-Finishing
Steer Calves and Yearlings.* (Dry Matter Basis)

Weight (lb)**	Daily Gain (lb)	Total Protein (lb)	Digestible Protein (lb)	TDN (lb)	Ca (g)	P (g)	Vit. A (thous. IU)
992	0	1.19	0.68	--	12	12	14
	2.2	2.11	1.25	16.3	20	20	20
	2.6	2.13	1.28	17.4	23	22	20
	2.9	2.13	1.30	17.6	24	23	20
	3.1	2.16	1.32	18.5	25	23	20
1,102	0	1.32	0.75	8.4	13	13	15
	2.0	2.09	1.23	16.5	19	19	23
	2.4	2.11	1.25	17.8	20	20	23
	2.6	2.11	1.28	18.1	21	21	23
	2.9	2.13	1.32	19.2	22	22	23

*Adapted from National Research Council (NRC), 1976.

**Weights taken from data in kilograms

Appendix Table III

Daily TDN, Protein, Ca, P and Vitamin A Requirements for Growing-Finishing Heifer Calves and Yearlings.* (Dry Matter Basis)

Weight (lb)**	Daily Gain (lb)	Total Protein (lb)	Digestible Protein (lb)	TDN (lb)	Ca (g)	P (g)	Vit. A (thous. IU)
220	0	0.40	0.22	2.6	4	4	5
	1.1	0.81	0.55	4.2	14	11	6
	1.5	0.92	0.64	4.6	19	14	6
	2.0	1.06	0.75	5.1	24	17	7
	2.4	1.17	0.86	5.5	29	19	7
331	0	0.53	0.31	3.5	5	5	6
	1.1	0.99	0.64	5.7	14	12	9
	1.5	1.10	0.73	6.2	18	14	9
	2.0	1.19	0.81	6.8	23	17	9
	2.4	1.32	0.92	7.5	28	20	9
441	0	0.66	0.37	4.2	6	6	8
	0.7	1.08	0.64	6.6	10	10	12
	1.1	1.28	0.77	7.7	14	13	13
	2.0	1.36	0.88	8.8	22	17	13
	2.4	1.41	0.97	9.5	25	19	13
551	0	0.77	0.44	5.1	7	7	9
	0.7	1.25	0.73	7.8	12	12	14
	1.1	1.36	0.81	8.6	13	13	14
	1.5	1.36	0.84	9.1	17	15	14
	2.0	1.43	0.92	10.1	21	17	14
	2.4	1.63	1.06	11.5	25	20	14
	2.6	1.65	1.08	11.9	27	21	14
661	0	0.88	0.51	5.7	9	9	10
	0.7	1.39	0.79	8.4	13	13	16
	1.1	1.47	0.88	9.9	14	14	16
	1.5	1.47	0.88	10.4	16	15	16
	2.0	1.54	0.97	11.5	19	17	16
	2.4	1.72	1.08	13.2	23	20	16
	2.6	1.74	1.10	13.7	24	20	16
772	0	1.01	0.57	6.4	10	10	12
	0.7	1.62	0.86	10.0	15	15	18
	1.1	1.61	0.92	11.2	15	15	18
	1.5	1.61	0.95	11.9	15	15	18
	2.0	1.69	1.01	13.2	17	17	18
	2.4	1.78	1.10	14.5	20	19	18
	2.6	1.78	1.10	15.2	21	20	18

Appendix Table III (continued)

Daily TDN, Protein, Ca, P and Vitamin A Requirements for Growing-Finishing
Heifer Calves and Yearlings.* (Dry Matter Basis)

Weight (lb)**	Daily Gain (lb)	Total Protein (lb)	Digestible Protein (lb)	TDN (lb)	Ca (g)	P (g)	Vit. A (thous. IU)
882	0	1.12	0.64	7.3	11	11	13
	0.7	1.67	0.95	11.1	16	16	19
	1.1	1.72	0.95	11.9	15	15	19
	1.5	1.74	1.01	13.2	16	16	19
	2.0	1.74	1.03	14.3	17	17	19
	2.4	1.78	1.08	15.9	19	18	19
992	0	1.21	0.68	7.9	12	12	14
	0.4	1.63	0.90	10.6	16	16	19
	1.1	1.76	1.01	13.0	17	17	20
	1.8	1.80	1.06	15.0	16	16	10
	2.2	1.83	1.06	16.3	19	19	20

*Adapted from NRC, 1976.

**Weights taken from data in kilograms

Appendix Table IV

Nutrient Requirements for Beef cattle Breeding Herd.* (Dry Matter Basis)

Weight (lb)**	Daily Gain (lb)	Total Protein (lb)	Digestible Protein (lb)	TDN (lb)	Ca (g)	P (g)	Vit. A (thous. IU)
Pregnant yearling heifers-Last 3-4 months of pregnancy							
716	0.9	1.3	0.7	7.7	15	15	19
	1.3	1.7	0.9	9.9	18	18	23
	1.8	1.9	1.1	12.3	22	20	26
772	0.9	1.3	0.8	8.1	15	15	19
	1.3	1.7	1.0	10.3	19	19	25
	1.8	1.9	1.1	12.9	22	21	28
827	0.9	1.4	0.8	8.4	15	15	20
	1.3	1.8	1.0	10.8	19	19	26
	1.8	2.1	1.2	13.5	22	22	31
882	0.9	1.4	0.8	8.7	16	16	21
	1.3	1.8	1.1	11.3	19	19	27
	1.8	2.2	1.3	14.0	22	22	33
937	0.9	1.5	0.9	9.0	16	16	22
	1.3	1.9	1.1	11.7	19	19	28
	1.8	2.3	1.3	14.6	22	22	34
Dry pregnant mature cows-Middle third of pregnancy							
772		0.7	0.3	6.6	10	10	15
882		0.8	0.4	7.3	11	11	17
992		0.9	0.4	7.9	12	12	19
1,102		0.9	0.4	8.6	13	13	20
1,213		1.0	0.5	9.2	14	14	22
1,323		1.1	0.5	9.8	15	15	23
1,433		1.1	0.6	10.4	16	16	25
Dry pregnant mature cows-Last third of pregnancy							
772	0.9	0.9	0.4	8.0	12	12	19
882	0.9	1.0	0.5	8.7	14	14	21
992	0.9	1.1	0.5	9.4	15	15	23
1,102	0.9	1.1	0.5	10.0	15	15	24
1,213	0.9	1.2	0.6	10.7	16	16	26
1,323	0.9	1.3	0.6	11.2	17	17	27
1,433	0.9	1.3	0.6	11.9	18	18	29

Appendix Table IV (continued)

Nutrient Requirements for Beef Cattle Breeding Herd.* (Dry Matter Basis)

Weight (lb)**	Daily Gain (lb)	Total Protein (lb)	Digestible Protein (lb)	TDN (lb)	Ca (g)	P (g)	Vit. A (thous. IU)
Cows nursing calves-Average milking ability[a]-First 3-4 months postpartum							
772		1.7	1.0	9.7	24	24	19
882		1.8	1.1	10.4	25	25	21
992		1.9	1.1	11.0	26	26	23
1,102		2.0	1.2	11.7	27	27	24
1,213		2.1	1.3	12.3	28	28	26
1,323		2.2	1.3	13.0	28	28	27
1,433		2.3	1.4	13.7	29	29	29
Cows nursing calves-Superior milking ability[b]-First 3-4 months postpartum							
772		2.4	1.4	12.8	45	40	32
882		2.6	1.5	13.5	45	41	34
992		2.7	1.6	14.1	45	42	36
1,102		2.8	1.7	14.8	46	43	38
1,213		3.0	1.7	15.4	46	44	43
1,323		3.1	1.8	16.1	46	44	43
1,433		3.2	1.9	16.8	47	45	45
Bulls, growth and maintenance (moderate activity)							
661	2.2	2.0	1.2	12.3	27	23	34
882	2.0	2.3	1.4	15.4	23	23	43
1,102	1.5	2.4	1.4	16.5	22	22	48
1,323	1.1	2.2	1.3	16.1	22	22	48
1.543	0.7	2.4	1.3	17.0	23	23	50
1,764	0	2.0	1.1	12.8	19	19	41
1,984	0	2.2	1.2	13.9	21	21	44
2,205	0	2.3	1.3	15.2	22	22	48

*Adapted from NRC, 1976.

**Weights taken from data in kilograms

[a]11.0 ± 1.1 lb milk/day

[b]22.0 ± 1.1 lb milk/day

Net Energy Requirements of Growing and Finishing Steers

Daily gain lb	\multicolumn{10}{c}{Body Weight lb}									
	350	360	370	380	390	400	410	420	430	440
	\multicolumn{10}{c}{NE_m required, megcal per day}									
0	3.48	3.55	3.63	3.70	3.77	3.85	3.92	3.99	4.06	4.13
	\multicolumn{10}{c}{NE_g required, megcal per day}									
0.5	0.55	0.56	0.57	0.59	0.60	0.61	0.62	0.63	0.64	0.65
0.6	0.67	0.68	0.69	0.71	0.72	0.74	0.75	0.76	0.78	0.79
0.7	0.78	0.80	0.81	0.83	0.85	0.86	0.88	0.89	0.91	0.93
0.8	0.90	0.92	0.94	0.95	0.97	0.99	1.01	1.03	1.05	1.06
0.9	1.01	1.04	1.06	1.08	1.10	1.12	1.14	1.16	1.18	1.20
1.0	1.13	1.16	1.18	1.21	1.23	1.25	1.28	1.30	1.32	1.35
1.1	1.25	1.28	1.31	1.33	1.36	1.39	1.41	1.44	1.46	1.49
1.2	1.38	1.40	1.43	1.46	1.49	1.52	1.55	1.58	1.60	1.63
1.3	1.50	1.53	1.56	1.59	1.62	1.66	1.69	1.72	1.75	1.78
1.4	1.62	1.66	1.69	1.73	1.76	1.79	1.83	1.86	1.89	1.93
1.5	1.75	1.78	1.82	1.86	1.90	1.93	1.97	2.00	2.04	2.07
1.6	1.87	1.91	1.95	1.99	2.03	2.07	2.11	2.15	2.19	2.22
1.7	2.00	2.04	2.09	2.13	2.17	2.21	2.25	2.29	2.34	2.38
1.8	2.13	2.18	2.22	2.27	2.31	2.35	2.40	2.44	2.49	2.53
1.9	2.26	2.31	2.36	2.40	2.45	2.50	2.55	2.59	2.64	2.68
2.0	2.39	2.44	2.49	2.54	2.59	2.64	2.69	2.74	2.79	2.84
2.1	2.52	2.58	2.63	2.69	2.74	2.79	2.84	2.89	2.95	3.00
2.2	2.66	2.72	2.77	2.83	2.88	2.94	2.99	3.05	3.10	3.16
2.3	2.79	2.85	2.91	2.97	3.03	3.09	3.15	3.20	3.26	3.32
2.4	2.93	2.99	3.05	3.12	3.18	3.24	3.30	3.36	3.42	3.48
2.5	3.07	3.13	3.20	3.26	3.33	3.39	3.45	3.52	3.58	3.64
2.6	3.21	3.28	3.34	3.41	3.48	3.54	3.61	3.68	3.74	3.81
2.7	3.35	3.42	3.49	3.56	3.63	3.70	3.77	3.84	3.91	3.97
2.8	3.49	3.56	3.64	3.71	3178	3.86	3.93	4.00	3.07	4.14
2.9	3.63	3.71	3.79	3.86	3.94	4.01	4.09	4.16	4.24	4.31
3.0	3.78	3.86	3.94	4.02	4.09	4.17	4.25	4.33	4.41	4.48
3.1	3.92	4.00	4.09	4.17	4.25	4.33	4.41	4.49	4.57	4.65
3.2	4.07	4.15	4.24	4.33	4.41	4.49	4.58	4.66	4.75	4.83
3.3	4.21	4.30	4.39	4.48	4.57	4.66	4.74	4.83	4.92	5.00
3.4	4.36	4.46	4.55	4.64	4.73	4.82	4.91	5.00	5.09	5.18

Net Energy Requirements of Growing and Finishing Steers (continued)

Daily gain lb	Body Weight lb									
	450	460	470	480	490	500	510	520	530	540
	NE_m required, megcal per day									
0	4.20	4.27	4.34	4.41	4.48	4.55	4.61	4.68	4.75	4.82
	NE_g required, megcal per day									
0.5	0.67	0.68	0.69	0.70	0.71	0.72	0.73	0.74	0.75	0.76
0.6	0.80	0.82	0.83	0.84	0.86	0.87	0.88	0.90	0.91	0.92
0.7	0.94	0.96	0.97	0.99	1.00	1.02	1.03	1.05	1.07	1.08
0.8	1.08	1.10	1.12	1.14	1.15	1.17	1.19	1.21	1.22	1.24
0.9	1.22	1.25	1.27	1.29	1.31	1.33	1.35	1.37	1.39	1.40
1.0	1.37	1.39	1.41	1.44	1.46	1.48	1.50	1.53	1.55	1.57
1.1	1.51	1.54	1.56	1.59	1.61	1.64	1.66	1.69	1.71	1.74
1.2	1.66	1.69	1.71	1.74	1.77	1.80	1.82	1.85	1.88	1.90
1.3	1.81	1.84	1.87	1.90	1.93	1.96	1.99	2.02	2.05	2.07
1.4	1.96	1.99	2.02	2.05	2.09	2.12	2.15	2.18	2.21	2.24
1.5	2.11	2.14	2.18	2.21	2.25	2.28	2.32	2.35	2.39	2.42
1.6	2.26	2.30	2.34	2.37	2.41	2.45	2.48	2.52	2.56	2.59
1.7	2.42	2.46	2.50	2.54	2.57	2.61	2.65	2.69	2.73	2.77
1.8	2.57	2.62	2.66	2.70	2.74	2.78	2.82	2.87	2.91	2.95
1.9	2.73	2.78	2.82	2.86	2.91	2.95	3.00	3.04	3.09	3.10
2.0	2.89	2.94	2.98	3.03	3.08	3.12	3.17	3.22	3.27	3.31
2.1	3.05	3.10	3.15	3.20	3.25	3.30	3.35	3.40	3.45	3.49
2.2	3.21	3.26	3.32	3.37	3.42	3.47	3.53	3.58	3.63	3.68
2.3	3.37	3.43	3.48	3.54	3.59	3.65	3.70	3.76	3.82	3.87
2.4	3.54	3.60	3.65	3.71	3.77	3.83	3.89	3.94	4.00	4.06
2.5	3.70	3.77	3.83	3.89	3.95	4.01	4.07	4.13	4.19	4.25
2.6	3.87	3.94	4.00	4.06	4.13	4.19	4.25	4.32	4.38	4.44
2.7	4.04	4.11	4.17	4.24	4.31	4.37	4.44	4.50	4.57	4.63
2.8	4.21	4.28	4.35	4.42	4.49	4.56	4.63	4.70	4.76	4.83
2.9	4.38	4.46	4.53	4.60	4.67	4.74	4.81	4.89	4.96	5.03
3.0	4.56	4.63	4.71	4.78	4.86	4.93	5.01	5.08	5.16	5.22
3.1	4.73	4.81	4.89	4.97	5.04	5.12	5.20	5.28	5.35	5.43
3.2	4.91	4.00	5.07	5.15	5.23	5.31	5.39	5.47	5.55	5.63
3.3	5.09	5.17	5.26	5.34	5.42	5.51	5.59	5.67	5.76	5.83
3.4	5.27	5.36	5.44	5.53	5.61	5.70	5.79	5.87	5.96	6.04

Net Energy Requirements of Growing and Finishing Steers (continued)

Body Weight lb

Daily gain lb	550	560	570	580	590	600	610	620	630	640
				NE_m required, megcal per day						
0	4.88	4.95	5.02	5.08	5.15	5.21	5.28	5.34	5.41	5.47
				NE_g required, megcal per day						
0.5	0.77	0.78	0.80	0.81	0.82	0.83	0.84	0.85	0.86	0.87
0.6	0.93	0.95	0.96	0.97	0.98	1.00	1.01	1.02	1.03	1.05
0.7	1.10	1.11	1.13	1.14	1.15	1.17	1.18	1.20	1.21	1.23
0.8	1.26	1.28	1.29	1.31	1.33	1.34	1.36	1.38	1.39	1.41
0.9	1.42	1.44	1.46	1.48	1.50	1.52	1.54	1.56	1.58	1.60
1.0	1.59	1.61	1.64	1.66	1.68	1.70	1.72	1.74	1.76	1.78
1.1	1.76	1.79	1.81	1.83	1.85	1.88	1.90	1.92	1.95	1.97
1.2	1.93	1.96	1.98	2.01	2.03	2.06	2.09	2.11	2.14	2.16
1.3	2.10	2.13	2.16	2.19	2.22	2.24	2.27	2.30	2.33	2.36
1.4	2.28	2.31	2.34	2.37	2.40	2.43	2.46	2.49	2.52	2.55
1.5	2.45	2.49	2.52	2.55	2.58	2.62	2.65	2.68	2.72	2.75
1.6	2.63	2.67	2.70	2.74	2.77	2.81	2.84	2.88	2.91	2.95
1.7	2.81	2.85	2.89	2.92	2.96	3.00	3.03	3.07	3.11	3.15
1.8	2.99	3.03	3.07	3.11	3.15	3.19	3.23	3.27	3.31	3.35
1.9	3.17	3.22	3.26	3.30	3.34	3.39	3.43	3.47	3.51	3.56
2.0	3.36	3.41	3.45	3.49	3.54	3.58	3.63	3.67	3.72	3.76
2.1	3.54	3.59	3.64	3.69	3.73	3.78	3.83	3.88	3.93	3.97
2.2	3.73	3.78	3.83	3.88	3.93	3.98	4.03	4.08	4.13	4.18
2.3	3.92	3.98	4.03	4.08	4.13	4.18	4.24	4.29	4.34	4.40
2.4	4.11	4.17	4.23	4.28	4.33	4.39	4.44	4.50	4.56	4.61
2.5	4.31	4.37	4.43	4.48	4.54	4.59	4.65	4.71	4.77	4.83
2.6	4.50	4.57	4.63	4.68	4.74	4.80	4.86	4.92	4.99	5.04
2.7	4.70	4.77	4.83	4.89	4.95	5.01	5.08	5.14	5.20	5.27
2.8	4.90	4.97	5.03	5.10	5.16	5.22	5.29	5.35	5.42	5.49
2.9	5.10	5.17	5.24	5.30	5.37	5.44	·4.41	5.57	5.65	5.71
3.0	5.30	5.37	5.45	5.51	5.58	5.65	5.72	5.79	5.87	5.94
3.1	5.50	5.58	5.65	5.73	5.80	5.87	5.94	6.02	6.09	6.17
3.2	5.71	5.79	5.87	5.94	6.02	6.09	6.17	6.24	6.32	6.40
3.3	5.92	6.00	6.08	6.16	6.23	6.31	6.39	6.47	6.55	6.63
3.4	6.13	6.21	6.29	6.37	6.45	6.54	6.62	6.70	6.78	6.86

Net Energy Requirements of Growing and Finishing Steers (continued)

Daily gain lb	Body Weight lb									
	650	660	670	680	690	700	710	720	730	740
	NE_m required, megcal per day									
0	5.53	5.60	5.66	5.73	5.79	5.85	5.91	5.98	6.04	6.10
	NE_g required, megcal per day									
0.5	0.88	0.89	0.90	0.91	0.92	0.93	0.94	0.95	0.96	0.97
0.6	1.06	1.07	1.08	1.09	1.11	1.12	1.13	1.14	1.15	1.17
0.7	1.24	1.26	1.27	1.28	1.30	1.31	1.33	1.34	1.35	1.37
0.8	1.43	1.44	1.46	1.48	1.49	1.51	1.52	1.54	1.56	1.57
0.9	1.61	1.63	1.65	1.67	1.69	1.71	1.72	1.74	1.76	1.78
1.0	1.80	1.82	1.85	1.87	1.89	1.91	1.93	1.95	1.97	1.99
1.1	1.99	2.02	2.04	2.06	2.09	2.11	2.13	2.15	2.18	2.20
1.2	2.19	2.21	2.24	2.26	2.29	2.31	2.34	2.36	2.39	2.41
1.3	2.38	2.41	2.44	2.47	2.49	2.52	2.55	2.57	2.60	2.63
1.4	2.58	2.61	2.64	2.67	2.70	2.73	2.76	2.79	2.81	2.84
1.5	2.78	2.81	2.84	2.88	2.91	2.94	2.97	3.00	3.03	3.06
1.6	2.98	3.01	3.05	3.08	3.12	3.15	3.18	3.22	3.25	3.29
1.7	3.18	3.22	3.26	3.29	3.33	3.37	3.40	3.44	3.47	3.51
1.8	3.39	3.43	3.47	3.51	3.54	3.58	3.62	3.66	3.70	3.74
1.9	3.60	3.64	3.68	3.72	3.76	3.80	3.84	3.88	3.92	3.96
2.0	3.80	3.85	3.89	3.94	3.98	4.02	4.06	4.11	4.15	4.19
2.1	4.02	4.06	4.11	4.16	4.20	4.25	4.29	4.34	4.38	4.43
2.2	4.23	4.28	4.33	4.38	4.42	4.47	4.52	4.57	4.61	4.66
2.3	4.44	4.49	4.55	4.60	4.65	4.70	4.75	4.80	4.85	4.90
2.4	4.66	4.71	4.77	4.82	4.87	4.93	4.98	5.03	5.08	5.14
2.5	4.88	4.93	4.99	5.05	5.10	5.16	5.21	5.27	5.32	5.38
2.6	5.10	5.16	5.22	5.28	5.33	5.39	5.45	5.51	5.56	5.62
2.7	5.32	5.38	5.45	5.51	5.57	5.63	5.60	5.75	5.81	5.87
2.8	5.55	5.61	5.68	5.74	5.80	5.87	5.93	5.99	6.05	6.12
2.9	5.78	5.84	5.91	5.98	6.04	6.11	6.17	6.24	6.30	6.37
3.0	6.00	6.07	6.14	6.21	6.28	6.35	6.41	6.49	6.55	6.62
3.1	6.24	6.30	6.38	6.45	6.52	6.59	6.66	6.73	6.80	6.87
3.2	6.47	6.54	6.62	6.69	6.76	6.84	6.91	6.99	7.06	7.13
3.3	6.70	6.78	6.86	6.94	7.01	7.09	7.16	7.24	7.31	7.39
3.4	6.94	7.02	7.10	7.18	7.26	7.34	7.41	7.50	7.57	7.65

Net Energy Requirements of Growing and Finishing Steers (continued)

Daily gain lb	Body Weight lb									
	750	760	770	780	790	800	810	820	830	840
	NE_m required, megcal per day									
0	6.16	6.23	6.29	6.35	6.41	6.47	6.53	6.59	6.65	6.71
	NE_g required, megcal per day									
0.5	0.98	0.99	1.00	1.01	1.02	1.02	1.03	1.04	1.05	1.06
0.6	1.18	1.19	1.20	1.21	1.22	1.24	1.25	1.26	1.27	1.28
0.7	1.38	1.40	1.41	1.42	1.44	1.45	1.46	1.48	1.49	1.50
0.8	1.59	1.61	1.62	1.64	1.65	1.67	1.68	1.70	1.71	1.73
0.9	1.80	1.82	1.83	1.85	1.87	1.89	1.90	1.92	1.94	1.96
1.0	2.01	2.03	2.05	2.07	2.09	2.11	2.13	21.5	2.17	2.19
1.1	2.22	2.24	2.27	2.29	2.31	2.33	2.35	2.37	2.40	2.42
1.2	2.44	2.46	2.48	2.52	2.53	2.56	2.58	2.60	2.63	2.65
1.3	2.65	2.68	2.71	2.73	2.76	2.78	2.81	2.84	2.86	2.89
1.4	2.87	2.90	2.93	2.96	2.99	3.01	3.04	3.07	3.10	3.13
1.5	3.09	3.13	3.16	3.19	3.22	3.25	3.28	3.31	3.34	3.37
1.6	3.32	3.35	3.39	3.42	3.45	3.48	3.51	3.55	3.58	3.61
1.7	3.54	3.58	3.62	3.65	3.69	3.72	3.75	3.79	3.82	3.86
1.8	3.77	3.81	3.85	3.89	3.92	3.96	4.00	4.03	4.07	4.11
1.9	4.00	4.05	4.08	4.12	4.16	4.20	4.24	4.28	4.32	4.36
2.0	4.23	4.28	4.32	4.36	4.40	4.45	4.49	4.53	4.57	4.61
2.1	4.47	4.52	4.56	4.61	4.65	4.69	4.74	4.78	4.82	4.87
2.2	4.71	4.76	4.80	4.85	4.90	4.94	4.99	5.03	5.08	5.13
2.3	4.95	5.00	5.05	5.10	5.14	5.19	1.24	5.29	5.34	5.39
2.4	5.19	5.24	5.29	5.35	5.40	5.45	5.50	5.55	5.60	5.65
2.5	5.43	5.49	5.54	5.60	5.65	5.70	5.76	5.81	5.86	5.92
2.6	5.68	5.74	5.79	5.85	5.91	5.96	6.02	6.07	6.13	6.18
2.7	5.93	5.99	6.05	6.11	6.16	6.22	6.28	6.34	6.39	6.45
2.8	6.18	6.24	6.30	6.36	6.42	6.48	6.54	6.60	6.67	6.73
2.9	6.43	6.50	6.56	6.62	6.69	6.75	6.81	6.87	6.94	7.00
3.0	6.69	6.76	6.82	6.89	6.95	7.02	7.08	7.15	7.21	7.28
3.1	6.94	7.02	7.08	7.15	7.22	7.29	7.35	7.42	7.49	7.56
3.2	7.20	7.28	7.35	7.42	7.49	7.56	7.63	7.70	7.77	7.64
3.3	7.46	7.54	7.61	7.69	7.76	7.83	7.91	7.98	8.05	8.13
3.4	7.73	7.81	7.88	7.96	8.04	8.11	8.19	8.26	8.34	8.41

Net Energy Requirements of Growing and Finishing Steers (continued)

Body Weight lb

Daily gain lb	850	860	870	880	890	900	910	920	930	940
				NE_m required, megcal per day						
0	6.77	6.83	6.89	6.95	7.00	7.06	7.13	7.18	7.24	7.30
				NE_g required, megcal per day						
0.5	1.07	1.08	1.09	1.10	1.11	1.12	1.13	1.14	1.15	1.16
0.6	1.29	1.31	1.32	1.33	1.34	1.35	1.36	1.37	1.38	1.40
0.7	1.52	1.53	1.55	1.56	1.57	1.58	1.60	1.61	1.62	1.64
0.8	1.75	1.76	1.78	1.79	1.81	1.82	1.84	1.85	1.87	1.88
0.9	1.97	1.99	2.01	2.03	2.04	2.06	2.08	2.09	2.11	2.13
1.0	2.21	2.22	2.24	2.26	2.28	2.30	2.32	2.34	2.36	2.38
1.1	2.44	2.46	2.48	2.50	2.52	2.55	2.57	2.59	2.61	2.63
1.2	2.68	2.70	2.72	2.75	2.77	2.79	2.82	2.84	2.86	2.89.
1.3	2.92	2.94	2.97	2.99	3.02	3.04	3.07	3.09	3.12	3.14
1.4	3.16	3.18	3.21	3.24	3.27	3.29	3.32	3.35	3.38	3.40
1.5	3.40	3.43	3.46	3.49	3.52	3.55	3.58	3.61	3.64	3.67
1.6	3.65	3.68	3.71	3.74	3.77	3.80	3.84	3.87	3.90	3.93
1.7	3.90	3.93	3.96	4.00	4.03	4.06	4.10	4.13	4.17	4.20
1.8	4.15	4.18	4.22	4.25	4.29	4.33	4.36	4.40	4.43	4.47
1.9	4.40	4.44	4.48	4.51	4.55	4.59	4.63	4.67	4.70	4.74
2.0	4.66	4.69	4.74	4.78	4.82	4.86	4.90	4.94	4.98	5.02
2.1	4.91	4.95	5.00	5.04	5.08	5.13	5.17	5.21	5.25	5.30
2.2	5.17	5.22	5.26	5.31	5.35	5.40	5.44	5.49	5.53	5.58
2.3	5.44	5.48	5.53	5.58	5.62	5.67	5.72	5.77	5.81	5.86
2.4	5.70	5.75	5.80	5.85	5.90	5.95	6.00	6.05	6.10	6.15
2.5	5.97	6.02	6.07	6.13	6.18	6.23	6.28	6.33	6.38	6.44
2.6	6.24	6.29	6.35	6.40	6.46	6.51	6.57	6.62	6.67	6.73
2.7	6.51	6.57	6.63	6.68	6.74	6.80	6.85	6.91	6.97	7.02
2.8	6.79	6.85	6.91	6.97	7.02	7.08	7.14	7.20	7.26	7.32
2.9	7.07	7.13	7.19	7.25	7.31	7.37	7.44	7.49	7.56	7.62
3.0	7.35	7.41	7.47	7.54	7.60	7.67	7.73	7.79	7.86	7.92
3.1	7.63	7.69	7.76	7.83	7.89	7.67	7.73	7.79	7.86	7.92
3.2	7.92	7.98	8.05	8.12	8.19	8.26	8.33	8.39	8.46	8.53
3.3	8.20	8.27	8.34	8.42	8.48	8.56	8.63	8.70	8.77	8.84
3.4	8.49	8.56	8.64	8.71	8.78	8.86	8.93	9.00	9.08	9.16

Net Energy Requirements of Growing and Finishing Steers (continued)

				Body Weight lb					
Daily gain lb	950	960	970	980	990	1000	1010	1020	1030	1040
				NE_m required, megcal per day					
0	7.36	7.42	7.47	7.53	7.59	7.65	7.71	7.76	7.82	7.87
				NE_g required, megcal per day					
0.5	1.17	1.18	1.18	1.19	1.20	1.21	1.22	1.23	1.24	1.25
0.6	1.41	1.42	1.43	1.44	1.45	1.46	1.47	1.48	1.49	1.51
0.7	1.65	1.66	1.68	1.69	1.70	1.71	1.73	1.74	1.75	1.77
0.8	1.90	1.91	1.93	1.94	1.96	1.97	1.99	2.00	2.02	2.03
0.9	2.15	2.16	2.18	2.20	2.21	2.23	2.25	2.26	2.28	2.30
1.0	2.40	2.42	2.43	2.45	2.47	2.49	2.51	2.53	2.55	2.57
1.1	2.65	2.67	2.69	2.71	2.73	2.76	2.78	2.80	2.82	2.84
1.2	2.91	2.93	2.95	2.98	3.00	3.02	3.05	3.07	3.09	3.11
1.3	3.17	3.19	3.22	3.24	3.27	3.29	3.32	3.34	3.37	3.39
1.4	3.43	3.46	3.48	3.51	3.54	3.56	3.59	3.62	3.64	3.67
1.5	3.69	3.72	3.75	3.78	3.81	3.84	3.87	3.90	3.93	3.95
1.6	3.96	3.99	4.02	4.06	4.09	4.12	4.15	4.18	4.21	4.24
1.7	4.23	4.27	4.30	4.33	4.37	4.40	4.43	4.46	4.50	4.53
1.8	4.50	4.54	4.58	4.61	4.65	4.68	4.72	4.75	4.79	4.82
1.9	4.78	4.82	4.86	4.89	4.93	4.97	5.01	5.04	5.08	5.12
2.0	5.06	5.10	5.14	5.18	5.22	5.26	5.30	5.34	5.37	5.41
2.1	5.34	5.38	5.42	5.47	5.51	5.55	5.59	5.63	5.67	5.71
2.2	5.62	5.67	5.71	5.76	5.80	5.84	5.89	5.93	5.97	6.02
2.3	5.91	5.96	6.00	6.05	6.09	6.14	6.19	6.23	6.28	6.32
2.4	6.20	6.25	6.29	6.34	6.39	6.44	6.49	6.54	6.58	6.63
2.5	6.49	6.54	6.59	6.64	6.69	6.74	6.79	6.84	6.89	6.94
2.6	6.78	6.84	6.89	6.94	7.00	7.05	7.10	7.15	7.20	7.26
2.7	7.08	7.14	7.19	7.25	7.30	7.35	7.41	7.47	7.52	7.57
2.8	7.38	7.44	7.49	7.55	7.61	7.67	7.73	7.78	7.84	7.89
2.9	7.68	7.74	7.80	7.86	7.92	7.98	8.04	8.10	8.16	8.22
3.0	7.98	8.05	8.11	8.17	8.23	8.30	8.36	8.42	8.48	8.54
3.1	8.29	8.36	8.42	8.49	8.55	8.61	8.68	8.74	8.81	8.87
3.2	8.60	8.67	8.73	8.81	8.87	8.94	9.01	9.07	9.17	9.20
3.3	8.91	9.98	9.05	9.13	9.19	9.26	9.33	9.40	9.47	9.54
3.4	9.23	9.30	9.37	9.45	9.52	9.59	9.66	9.73	9.80	9.87

Net Energy Requirements of Growing and Finishing Steers (continued)

Daily gain lb	Body Weight lb					
	1050	1060	1070	1080	1090	1100
0	NE_m required, megcal per day					
0	7.93	7.99	8.05	8.10	8.16	8.21
	NE_g required, megcal per day					
0.5	1.26	1.27	1.27	1.28	1.29	1.30
0.6	1.52	1.53	1.54	1.55	1.56	1.57
0.7	1.78	1.79	1.80	1.82	1.83	1.84
0.8	2.05	2.06	2.07	2.09	2.10	2.12
0.9	2.31	2.33	2.35	2.36	2.38	2.39
1.0	2.58	2.60	2.62	2.64	2.66	2.68
1.1	2.86	2.88	2.90	2.92	2.94	2.96
1.2	3.14	3.16	3.18	3.20	3.22	3.25
1.3	3.42	3.44	3.46	3.49	3.51	3.54
1.4	3.70	3.72	3.75	3.78	3.80	3.83
1.5	3.98	4.01	4.04	4.07	4.10	4.12
1.6	4.27	4.30	4.33	4.36	4.39	4.42
1.7	4.56	4.60	4.63	4.66	4.69	4.72
1.8	4.86	4.89	4.93	4.96	4.99	5.03
1.9	5.15	5.19	5.23	5.26	5.30	5.34
2.0	5.45	5.49	5.53	5.57	5.61	5.65
2.1	5.76	5.80	5.84	5.88	5.92	5.96
2.2	6.06	6.10	6.15	6.19	6.23	6.28
2.3	6.37	6.42	6.46	6.50	6.55	6.59
2.4	6.68	6.73	6.78	6.82	6.87	6.92
2.5	7.00	7.04	7.09	7.14	7.19	7.24
2.6	7.31	7.36	7.41	7.47	7.52	7.57
2.7	7.63	7.69	7.74	7.79	7.85	7.90
2.8	7.95	8.01	8.07	8.12	8.18	8.23
2.9	8.28	8.34	8.40	8.45	8.51	8.57
3.0	8.61	8.67	8.73	8.79	8.85	8.91
3.1	8.94	9.00	9.06	9.13	9.19	9.25
3.2	9.27	9.34	9.40	9.47	9.53	9.60
3.3	9.61	9.68	9.74	9.81	9.88	9.95
3.4	9.95	10.02	10.09	10.16	10.23	10.30

Net Energy Requirements of Growing and Finishing Heifers

Daily gain lb	Body Weight lb									
	350	360	370	380	390	400	410	420	430	440
0				NE_m required, megcal per day						
0	3.48	3.55	3.63	3.70	3.77	3.85	3.92	3.99	4.06	4.13
				NE_g required, megcal per day						
0.5	0.60	0.16	0.62	0.64	0.65	0.66	0.67	0.69	0.70	0.71
0.6	0.73	0.74	0.76	0.77	0.79	0.80	0.82	0.83	0.85	0.86
0.7	0.85	0.87	0.89	0.91	0.93	0.94	0.96	0.98	1.00	1.01
0.8	0.99	1.01	1.03	1.05	1.07	1.09	1.11	1.13	1.15	1.17
0.9	1.12	1.14	1.17	1.19	1.21	1.24	1.26	1.28	1.31	1.33
1.0	1.26	1.28	1.31	1.33	1.36	1.39	1.41	1.44	1.46	1.49
1.1	1.39	1.42	1.45	1.48	1.51	1.54	1.47	1.60	1.63	1.65
1.2	1.53	1.57	1.60	1.63	1.66	1.69	1.73	1.76	1.79	1.82
1.3	1.68	1.71	1.75	1.78	1.82	1.85	1.89	1.92	1.96	1.99
1.4	1.82	1.86	1.90	1.94	1.98	2.01	2.05	2.09	2.13	2.16
1.5	1.97	2.01	2.05	2.09	2.14	2.18	2.22	2.26	2.30	2.34
1.6	2.12	2.16	2.21	2.25	2.30	2.34	2.39	2.43	2.47	2.52
1.7	2.27	2.32	2.37	2.42	2.46	2.51	2.56	2.60	2.65	2.70
1.8	2.43	2.48	2.53	2.58	2.63	2.68	2.73	2.78	2.83	2.88
1.9	2.58	2.64	2.69	2.75	2.80	2.85	2.91	2.96	3.01	3.07
2.0	2.74	2.80	2.86	2.92	2.97	3.03	3.09	3.14	3.20	3.25
2.1	2.90	2.96	3.03	3.09	3.15	3.21	3.27	3.33	3.39	3.45
2.2	3.07	3.13	3.20	3.26	3.32	3.39	3.45	3.52	3.58	3.64
2.3	3.23	3.30	3.37	3.44	3.50	3.57	3.64	3.71	3.77	3.84
2.4	3.40	3.47	3.54	3.62	3.69	3.96	3.83	3.90	3.97	4.04
2.5	3.57	3.65	3.72	3.80	3.87	3.95	4.02	4.09	4.17	4.24
2.6	3.74	3.82	3.90	3.98	4.06	4.14	4.21	4.29	4.37	4.44
2.7	3.92	4.00	4.09	4.17	4.25	4.33	4.41	4.49	4.57	4.65
2.8	4.10	4.18	4.27	4.36	4.44	4.53	4.61	4.70	4.78	4.86
2.9	4.28	4.37	4.46	4.55	4.64	4.73	4.81	4.90	4.99	5.08
3.0	4.46	4.55	4.65	4.74	4.83	4.93	50.2	5.11	5.20	5.29
3.1	4.64	4.74	4.84	4.94	5.03	5.13	5.23	5.32	5.42	5.51
3.2	4.83	4.93	5.03	5.14	5.24	5.34	5.44	5.54	5.64	5.73
3.3	5.02	5.12	5.23	5.34	5.44	5.55	5.65	5.75	5.86	5.96
3.4	5.21	5.32	5.43	5.54	5.65	5.76	5.87	5.97	6.08	6.19

Net Energy Requirements of Growing and Finishing Heifers (continued)

Body Weight lb

Daily gain lb	450	460	470	480	490	500	510	· 520	530	540
	NE$_m$ required, megcal per day									
0	4.20	4.27	4.34	4.41	4.48	4.55	4.61	4.68	4.75	4.82
	NE$_g$ required, megcal per day									
0.5	0.72	0.73	0.75	0.76	0.77	0.78	0.79	0.81	0.82	0.83
0.6	0.88	0.89	0.90	0.92	0.93	0.95	0.96	0.98	0.99	1.00
0.7	1.03	1.05	1.07	1.08	1.10	1.12	1.13	1.15	1.17	1.18
0.8	1.19	1.21	1.23	1.25	1.27	1.29	1.31	1.33	1.35	1.36
0.9	1.35	1.37	1.40	1.42	1.44	1.46	1.48	1.51	1.53	1.55
1.0	1.52	1.54	1.56	1.59	1.61	1.64	1.66	1.69	1.71	1.74
1.1	1.68	1.71	1.74	1.76	1.79	1.82	1.85	1.87	1.90	1.93
1.2	1.85	1.88	1.91	1.94	1.97	2.00	2.03	2.06	2.09	2.12
1.3	2.02	2.06	2.09	2.12	2.16	2.19	2.22	2.26	2.29	2.32
1.4	2.20	2.24	2.27	2.31	2.34	2.38	2.42	2.45	2.49	2.52
1.5	2.38	2.42	2.46	2.49	2.53	2.57	2.61	2.65	2.69	2.73
1.6	2.56	2.60	2.64	2.68	2.73	2.77	2.81	2.85	2.89	2.93
1.7	2.74	2.79	2.83	2.88	2.92	2.97	3.01	3.06	3.10	3.14
1.8	2.93	2.98	3.02	3.07	3.12	3.17	3.22	3.26	3.31	3.36
1.9	3.12	3.17	3.22	3.27	3.32	3.37	3.42	3.47	3.53	3.57
2.0	3.31	3.36	3.42	3.47	3.53	3.58	3.63	3.69	3.74	3.79
2.1	3.50	3.56	3.62	3.68	3.73	3.79	3.85	3.91	3.96	4.02
2.2	3.70	3.76	3.82	3.88	3.94	4.01	4.07	4.13	4.19	4.24
2.3	3.90	3.97	4.03	4.09	4.16	1.22	4.29	4.35	4.41	4.47
2.4	4.10	4.17	4.24	4.31	4.37	4.44	4.51	4.58	4.64	4.71
2.5	4.31	4.38	4.45	4.52	4.59	4.66	4.73	4.81	4.88	4.94
2.6	4.52	4.60	4.67	4.74	4.82	4.89	4.96	5.04	5.11	5.18
2.7	4.73	4.81	4.89	4.96	5.04	5.12	5.20	5.27	5.35	5.42
2.8	4.95	5.03	5.11	5.19	5.27	2.35	5.43	5.51	5.59	5.67
2.9	5.16	5.25	2.33	5.42	5.50	5.59	5.67	5.75	5.84	5.92
3.0	5.38	5.47	5.56	5.65	5.74	5.82	5.91	6.00	6.09	6.17
3.1	5.60	5.70	5.79	5.88	5.97	6.06	6.16	6.25	6.34	6.43
3.2	5.83	5.93	6.02	6.12	6.21	6.31	6.40	6.50	6.59	6.68
3.3	6.06	6.16	6.26	6.36	6.46	6.56	6.65	6.75	6.85	6.95
3.4	6.29	6.40	6.50	6.60	6.70	6.81	6.91	7.01	7.11	7.21

Net Energy Requirements of Growing and Finishing Heifers (continued)

Body Weight lb

Daily gain lb	550	560	570	580	590	600	610	620	630	640
	NE_m required, megcal per day									
0	4.88	4.95	5.02	5.08	5.15	5.21	5.28	5.34	5.41	5.47
	NE_g required, megcal per day									
0.5	0.84	0.85	0.86	0.87	0.89	0.90	0.91	0.92	0.93	0.94
0.6	1.02	1.03	1.05	1.06	1.07	1.09	1.10	1.11	1.13	1.14
0.7	1.20	1.22	1.23	1.25	1.26	1.28	1.30	1.31	1.33	1.34
0.8	1.38	1.40	1.42	1.44	1.46	1.48	1.49	1.51	1.53	1.55
0.9	1.57	1.59	1.61	1.63	1.66	1.68	1.70	1.72	1.74	1.76
1.0	1.76	1.79	1.81	1.83	1.86	1.88	1.90	1.93	1.95	1.97
1.1	1.96	1.98	2.01	2.03	2.06	2.09	2.11	2.14	2.17	2.19
1.2	2.15	2.18	2.21	2.24	2.27	2.30	2.33	2.35	2.38	2.41
1.3	2.35	2.39	2.42	2.45	2.48	2.51	2.54	2.57	2.61	2.64
1.4	2.56	2.59	2.63	2.66	2.69	2.73	2.76	2.80	2.83	2.87
1.5	2.76	2.80	2.84	2.88	2.91	2.95	2.99	3.02	3.06	3.10
1.6	2.97	3.02	3.06	3.09	3.13	3.17	3.21	3.25	3.29	3.33
1.7	3.19	3.23	3.27	3.32	3.36	3.40	3.44	3.49	3.53	3.57
1.8	3.40	3.45	3.50	3.54	3.59	3.63	3.68	3.72	3.77	3.82
1.9	3.62	3.68	3.72	3.77	3.82	3.87	3.91	3.96	4.01	4.06
2.0	3.85	3.90	3.95	4.00	4.05	4.11	4.16	4.21	4.26	4.31
2.1	4.07	4.13	4.19	4.24	4.29	4.35	4.40	4.45	4.51	4.57
2.2	4.30	4.36	4.42	4.48	4.54	4.59	4.65	4.71	4.77	4.82
2.3	4.54	4.60	4.66	4.72	4.78	4.84	4.90	4.96	5.02	5.08
2.4	4.77	4.84	4.90	4.97	5.03	5.09	5.16	5.22	5.29	5.35
2.5	5.01	5.08	5.15	5.22	5.28	5.35	5.41	5.48	5.55	5.62
2.6	5.26	5.33	5.40	5.47	5.54	5.61	5.68	5.75	5.82	5.89
2.7	5.50	5.58	5.65	5.72	5.80	5.87	5.94	6.01	6.09	6.16
2.8	5.75	5.83	5.91	5.98	6.06	6.14	6.21	6.29	6.37	6.44
2.9	6.00	6.09	6.17	6.25	6.33	6.40	6.48	6.56	6.65	6.73
3.0	6.36	6.45	6.43	6.51	6.60	6.68	6.76	6.84	6.93	7.01
3.1	6.52	6.61	6.70	6.78	6.87	6.95	7.04	7.13	7.22	7.30
3.2	6.78	6.87	6.96	7.05	7.14	7.23	7.32	7.41	7.51	7.60
3.3	7.04	7.14	7.24	7.33	7.42	7.52	7.61	7.70	7.80	7.89
3.4	7.31	7.42	7.51	7.61	7.71	7.80	7.90	8.00	8.10	8.20

Net Energy Requirements of Growing and Finishing Heifers (continued)

Daily gain lb	\multicolumn{10}{c}{Body Weight lb}
	650	660	670	680	690	700	170	720	730	740

NE_m required, megcal per day

Daily gain lb	650	660	670	680	690	700	170	720	730	740
0	5.53	5.60	5.66	5.73	5.79	5.85	5.91	5.98	6.04	6.10

NE_g required, megcal per day

Daily gain lb	650	660	670	680	690	700	170	720	730	740
0.5	0.95	0.96	0.97	0.99	1.00	1.01	1.02	1.03	1.04	1.05
0.6	1.15	1.17	1.18	1.19	1.21	1.22	1.23	1.25	1.26	1.27
0.7	1.36	1.37	1.39	1.41	1.42	1.44	1.45	1.47	1.48	1.50
0.8	1.57	1.59	1.60	1.62	1.64	1.66	1.67	1.69	1.71	1.73
0.9	1.78	1.80	1.82	1.84	1.86	1.88	1.90	1.92	1.94	1.96
1.0	2.00	2.02	2.04	2.07	2.09	2.11	2.13	2.16	2.18	2.20
1.1	2.22	2.24	2.27	2.29	2.32	2.34	2.37	2.39	2.42	2.44
1.2	2.44	2.47	2.50	2.52	2.55	2.58	2.61	2.63	2.66	2.69
1.3	2.67	2.70	2.73	2.76	2.79	2.82	2.85	2.88	2.91	2.94
1.4	2.90	2.93	2.96	3.00	3.03	3.06	3.10	3.13	1.36	1.39
1.5	3.13	3.17	3.20	3.24	3.28	3.31	3.35	3.38	3.42	3.45
1.6	3.37	3.41	3.45	3.49	3.52	3.56	3.60	3.64	3.68	3.72
1.7	3.61	3.65	3.70	3.74	3.78	3.82	3.86	3.90	3.94	3.98
1.8	3.86	3.90	3.95	3.99	4.03	4.08	4.12	4.17	4.21	4.25
1.9	4.11	4.15	4.20	4.25	4.29	4.34	4.39	4.43	4.48	4.53
2.0	4.36	4.41	4.46	4.51	4.56	4.61	4.66	4.71	4.76	4.81
2.1	4.62	4.67	4.72	4.78	4.83	4.88	4.93	4.99	5.04	5.09
2.2	4.88	4.93	4.99	5.05	5.10	5.16	5.21	5.27	5.32	5.38
2.3	5.14	5.20	5.26	5.32	5.38	5.44	5.49	5.55	5.61	5.67
2.4	5.41	5.47	5.53	5.60	5.66	5.72	5.78	5.84	5.90	5.96
2.5	5.68	5.75	5.81	5.88	5.94	6.01	6.07	6.13	6.20	6.26
2.6	5.95	6.02	6.09	6.16	6.23	6.30	6.36	6.43	6.49	6.57
2.7	6.23	6.31	6.38	6.45	6.52	6.59	6.66	6.73	6.80	6.87
2.8	6.51	6.59	6.67	6.74	6.81	6.89	6.96	7.04	7.11	7.18
2.9	6.80	6.88	6.96	7.04	7.11	7.19	7.27	7.34	7.42	7.50
3.0	7.09	7.17	7.26	7.34	7.42	7.50	7.58	7.66	7.73	7.82
3.1	7.38	7.47	7.56	7.64	7.72	7.81	7.89	7.97	8.05	8.14
3.2	7.68	7.77	7.86	7.95	8.03	8.12	8.21	8.30	8.38	8.47
3.3	7.98	8.08	8.17	8.26	8.35	8.44	8.53	8.62	8.71	8.80
3.4	8.29	8.38	8.48	8.58	8.67	8.76	8.85	8.95	9.04	9.14

Net Energy Requirements of Growing and Finishing Heifers (continued)

				Body Weight lb						
Daily gain lb	750	760	770	780	790	800	810	820	830	840
				NE_m required, megcal per day						
0	6.16	6.23	6.29	6.35	6.41	6.47	6.53	6.59	6.65	6.71
				NE_g required, megcal per day						
0.5	1.06	1.07	1.08	1.09	1.10	1.11	1.12	1.13	1.14	1.15
0.6	1.28	1.30	1.31	1.32	1.34	1.35	1.36	1.37	1.39	1.40
0.7	1.51	1.53	1.54	1.56	1.57	1.59	1.60	1.62	1.63	1.65
0.8	1.75	1.76	1.78	1.80	1.81	1.83	1.85	1.87	1.89	1.90
0.9	1.98	2.00	2.02	2.04	2.06	2.08	2.10	2.12	2.14	2.16
1.0	2.22	2.25	2.27	2.29	2.31	2.33	2.35	2.38	2.40	2.42
1.1	2.47	2.49	2.52	2.54	2.57	2.59	2.61	2.64	2.66	2.69
1.2	2.72	2.74	2.77	2.80	2.82	2.85	2.88	2.90	2.93	2.96
1.3	2.97	3.00	3.03	3.06	3.09	3.12	3.14	3.17	3.20	3.23
1.4	3.23	3.26	3.29	3.32	3.35	3.39	3.42	3.45	3.48	3.51
1.5	3.49	3.52	3.56	3.59	3.63	3.66	3.69	3.73	3.76	3.80
1.6	3.75	3.79	3.83	3.86	3.90	3.94	3.97	4.01	4.05	4.08
1.7	4.02	4.06	4.10	4.14	4.18	4.22	4.26	4.30	4.34	4.38
1.8	4.30	4.34	4.38	4.42	4.47	4.51	4.55	4.59	4.63	4.68
1.9	4.57	4.62	4.66	4.71	4.75	4.80	4.84	4.89	4.93	4.98
2.0	4.85	4.90	4.95	5.00	5.05	5.09	5.14	5.19	5.24	5.28
2.1	4.14	5.19	5.24	5.29	5.34	5.39	5.44	5.49	5.54	5.60
2.2	5.43	5.49	5.54	5.59	5.65	5.70	5.75	5.80	5.86	5.91
2.3	5.72	5.78	5.84	5.89	5.95	6.01	6.06	6.12	6.17	6.23
2.4	6.02	6.08	6.14	6.20	6.26	6.32	6.38	6.44	6.50	6.55
2.5	6.32	6.39	6.45	6.51	6.57	6.64	6.70	6.76	6.82	6.88
2.6	6.63	6.70	6.76	6.83	6.89	6.96	7.02	7.09	7.15	7.22
2.7	6.94	7.01	7.08	7.15	7.22	7.28	7.35	7.42	7.49	7.55
2.8	7.25	7.33	7.40	7.47	7.54	7.61	7.68	7.76	7.83	7.90
2.9	7.57	7.65	7.73	7.80	7.87	7.95	8.02	8.10	8.17	8.24
3.0	7.90	7.98	8.05	8.13	8.21	8.29	8.36	8.44	8.52	8.59
3.1	8.22	8.31	8.39	8.47	8.55	8.63	8.71	8.79	8.87	8.95
3.2	8.55	8.64	8.73	8.81	8.89	8.98	9.06	9.14	9.23	9.31
3.3	8.89	8.98	9.07	9.15	9.24	9.33	9.41	9.50	9.59	9.67
3.4	9.23	9.32	9.41	9.50	9.59	9.68	9.77	9.86	9.55	10.04

Net Energy Requirements of Growing and Finishing Heifers (continued)

Body Weight lb

Daily gain lb	850	860	870	880	890	900	910	920	930	940
	NE_m required, megcal per day									
0	6.77	6.83	6.89	6.95	7.00	7.06	7.13	7.18	7.24	7.30
	NE_g required, megcal per day									
0.5	1.17	1.17	1.19	1.20	1.20	1.22	1.23	1.24	1.25	1.26
0.6	1.41	1.42	1.44	1.45	1.46	1.47	1.49	1.50	1.51	1.52
0.7	1.66	1.68	1.69	1.71	1.72	1.73	1.75	1.76	1.78	1.79
0.8	1.92	1.93	1.95	1.97	1.98	2.00	2.02	2.03	2.05	2.07
0.9	2.18	2.20	2.22	2.23	2.25	2.27	2.29	2.31	2.33	2.35
1.0	2.44	2.46	2.48	2.51	2.53	2.55	2.57	2.59	2.61	2.63
1.1	2.71	2.73	2.76	2.78	2.80	2.83	2.85	2.88	2.90	2.92
1.2	2.98	3.01	3.04	3.06	3.09	3.11	3.14	3.17	3.19	3.22
1.3	3.26	3.29	2.32	3.35	3.37	3.40	3.43	3.46	3.49	3.52
1.4	3.55	3.57	3.61	3.64	3.67	3.70	3.73	3.76	3.79	3.82
1.5	3.83	3.86	3.90	3.93	3.96	4.00	4.03	4.06	4.10	4.13
1.6	4.12	4.16	4.19	4.23	4.27	4.30	4.34	4.37	4.41	4.45
1.7	4.42	4.46	4.50	4.54	4.57	4.61	4.65	4.69	4.73	4.77
1.8	4.72	4.76	4.80	4.84	4.88	4.92	4.97	5.01	5.05	5.09
1.9	5.03	5.07	5.11	5.16	5.20	5.24	5.29	5.33	5.37	5.42
2.0	5.33	5.38	5.43	5.47	5.52	5.57	5.61	5.66	5.70	5.75
2.1	5.65	5.70	5.75	5.80	5.84	5.89	5.94	5.99	6.04	6.09
2.2	5.97	6.02	6.07	6.12	6.17	6.23	6.28	6.33	6.38	6.43
2.3	6.29	6.34	6.40	6.45	6.51	6.56	6.62	6.67	6.73	6.78
2.4	6.62	6.67	6.73	6.79	6.84	6.90	6.96	7.02	7.08	7.13
2.5	6.95	7.01	7.07	7.13	7.19	7.25	7.31	7.37	7.43	7.49
2.6	7.29	7.35	7.41	7.48	7.54	7.60	7.67	7.73	7.79	7.86
2.7	7.63	7.69	7.76	7.83	7.89	7.96	8.02	8.09	8.16	8.22
2.8	7.97	8.04	8.11	8.18	8.25	8.32	8.39	8.45	8.52	8.60
2.9	8.32	8.39	8.47	8.54	8.61	8.68	8.76	8.82	8.90	8.97
3.0	8.68	8.75	8.83	8.90	8.98	9.05	9.13	9.20	9.28	9.35
3.1	9.04	9.11	9.19	9.27	9.35	9.43	9.51	9.58	9.66	9.74
3.2	9.40	9.48	9.56	9.64	9.72	9.81	9.89	9.97	10.05	10.13
3.3	9.77	9.85	9.94	10.02	10.10	10.19	10.28	10.36	10.44	10.53
3.4	10.14	10.22	10.31	10.40	10.49	10.58	10.67	10.75	10.84	10.93

Net Energy Requirements of Growing and Finishing Heifers (continued)

Daily gain lb	\multicolumn{10}{c}{Body Weight lb}									
	950	960	970	980	990	1000	1010	1020	1030	1040
	\multicolumn{10}{c}{NE_m required, megcal per day}									
0	7.36	7.42	7.47	7.53	7.59	7.65	7.71	7.76	7.82	7.87
	\multicolumn{10}{c}{NE_g required, megcal per day}									
0.5	1.27	1.28	1.29	1.30	1.31	1.32	1.33	1.34	1.34	1.35
0.6	1.53	1.55	1.56	1.57	1.58	1.59	1.61	1.62	1.63	1.64
0.7	1.81	1.82	1.83	1.85	1.86	1.88	1.89	1.91	1.92	1.93
0.8	2.08	2.10	2.12	2.13	2.15	2.17	2.18	2.20	2.21	2.23
0.9	2.37	2.39	2.40	2.42	2.44	2.46	2.48	2.50	2.51	2.53
1.0	2.65	2.68	2.70	2.72	2.74	2.76	2.78	2.80	2.82	2.84
1.1	2.95	2.97	3.00	3.02	3.04	3.06	3.09	3.11	3.13	3.15
1.2	3.24	3.27	3.29	3.32	3.35	3.37	3.40	3.42	3.45	3.47
1.3	3.54	3.57	3.60	3.63	3.66	3.68	3.71	3.74	3.77	3.79
1.4	3.85	3.88	3.91	3.94	3.97	4.00	4.03	4.06	4.09	4.12
1.5	4.16	4.20	4.23	4.26	4.29	2.33	4.36	4.39	4.42	4.46
1.6	4.48	4.52	4.55	4.59	4.62	4.66	4.69	4.73	4.76	4.79
1.7	4.80	4.84	4.88	4.92	4.95	4.99	5.03	5.07	5.10	5.14
1.8	5.13	5.17	5.21	5.25	5.29	5.33	5.37	5.41	5.45	5.49
1.9	5.46	5.50	5.55	5.59	5.63	5.67	5.72	5.76	5.80	5.84
2.0	5.80	5.84	5.89	5.93	5.98	6.02	6.07	6.11	6.16	6.20
2.1	6.14	6.19	6.13	6.28	6.33	6.38	6.43	6.47	6.52	6.57
2.2	6.48	6.54	6.58	6.64	6.69	6.74	6.79	6.84	6.89	6.94
2.3	6.83	6.89	6.94	7.00	7.05	7.10	7.16	7.21	7.26	7.31
2.4	7.19	7.25	7.30	7.36	7.42	7.47	7.53	7.58	7.64	7.69
2.5	7.55	7.61	7.67	7.73	7.79	7.85	7.91	7.96	8.02	8.08
2.6	7.92	7.98	8.04	8.10	8.17	8.23	8.29	8.35	8.41	8.47
2.7	8.29	8.35	8.42	8.48	8.55	8.61	8.68	8.74	8.80	8.87
2.8	8.66	8.73	8.80	8.87	8.93	9.00	9.07	9.14	9.20	9.27
2.9	9.04	9.12	9.18	9.26	9.33	9.40	9.47	9.54	9.61	9.68
3.0	9.43	9.50	9.58	9.65	9.72	9.80	9.87	9.94	10.02	10.09
3.1	9.82	9.90	9.97	10.05	10.13	10.20	10.28	10.36	10.43	10.51
3.2	10.21	10.30	10.37	10.46	10.53	10.61	10.69	10.77	10.85	10.93
3.3	10.61	10.70	10.78	10.87	10.95	11.03	11.11	11.19	11.28	11.36
3.4	11.02	11.11	11.19	11.28	11.36	11.45	11.54	11.62	11.70	11.79

All Cattle and Calves: Number on Farms and Ranches by States, Jan. 1, 1978-80

State	1978 (1,000 head)	1979 (1,000 head)	1980¹/ (1,000 head)	1980 as % of 79
Alabama	2,130	1,820	1,730	95
Alaska	8.3	8.5	8.4	99
Arizona	1,135	1,200	1,050	88
Arkansas	2,120	2,000	2,000	100
California	4,430	4,700	4,550	97
Colorado	3,180	3,090	2,975	96
Connecticut	108	101	104	103
Delaware	31	30	30	100
Florida	2,350	2,180	2,300	106
Georgia	1,975	1,650	1,600	97
Hawaii	234	215	213	99
Idaho	1,870	1,900	1,860	98
Illinois	2,950	2,850	2,700	95
Indiana	2,025	1,750	1,850	106
Iowa	7,800	7,300	7,150	98
Kansas	6,000	6,200	6,100	98
Kentucky	3,120	2,600	2,700	104
Louisiana	1,425	1,350	1,300	96
Maine	132	124	131	106
Maryland	390	370	380	103
Massachusetts	99	95	103	108
Michigan	1,470	1,250	1,310	105
Minnesota	3,700	3,650	3,750	103
Mississippi	2,130	1,790	1,810	101
Missouri	6,000	5,550	5,350	96
Montana	2,680	2,607	2,645	101
Nebraska	6,500	6,450	6,400	99
Nevada	570	560	580	104
New Hampshire	74	68	70	103
New Jersey	114	108	100	93
New Mexico	1,550	1,500	1,600	107
New York	1,760	1,711	1,780	104
North Carolina	1,100	1,080	1,080	100
North Dakota	2,050	1,967	2,000	102
Ohio	2,025	1,850	1,925	104
Oklahoma	5,900	5,300	5,500	104
Oregon	1,490	1,475	1,510	102
Pennsylvania	1,900	1,840	1,900	103
Rhode Island	10.0	9.0	8.0	89
South Carolina	690	575	625	109
South Dakota	3,925	3,750	4,010	107
Tennessee	2,700	2,350	2,300	98
Texas	14,500	13,900	13,200	95
Utah	864	810	840	104
Vermont	336	320	340	106
Virginia	1,620	1,550	1,750	113
Washington	1,275	1,375	1,579	115
West Virginia	550	535	545	102
Wisconsin	4,100	4,100	4,280	104
Wyoming	1,280	1,300	1,340	103
TOTAL	116,375	110,864	110,961	100

1/Preliminary

Cattle, Sheep and Hogs on Farms Jan. 1, 1925-80
Also Total U. S. Resident Population July 1, 1925-79

YEAR	CATTLE AND CALVES (1,000 head)	SHEEP AND LAMBS (1,000 head)	TOTAL HOGS 2/ (1,000 head)	RESIDENT POPULATION (Millions)
1925	63,373	38,543	55,570	115.8
1930	61,003	51,565	55,795	123.1
1935	68,845	51,808	39,066	127.3
1940	68,309	46,266	61,165	132.0
1945	85,573	39,609	59,373	132.5
1950	77,963	29,826	58,937	151.9
1951	82,083	30,633	62,269	154.0
1952	88,072	31,982	62,117	156.4
1953	94,241	31,900	51,755	159.0
1954	95,679	31,356	45,114	161.9
1955	96,592	31,582	50,474	165.1
1956	95,900	31,157	55,354	168.1
1957	92,860	30,654	51,897	171.2
1958	91,176	31,217	51,517	174.1
1959	93,332	32,606	58,045	177.1
1960	96,236	33,170	59,026	179.3
1961	97,700	32,725	55,506	182.3
1962	100,369	30,969	57,000	185.1
1963	104,488	29,176	62,061	187.8
1964	107,903	27,116	56,106	190.5
1965	109,000	25,127	50,519	193.0
1966	108,862	24,734	57,125	195.0
1967	108,783	23,953	58,818	197.0
1968	109,371	22,223	60,829	198.9
1969	110,015	21,350	57,046	200.9
1970	112,369	20,423	67,285	203.2
1971	114,578	19,731	62,412	205.7
1972	117,862	18,739	51,017	207.8
1973	121,539	17,641	60,614	209.5
1974	127,788	16,310	54,693	209.7
1975	132,028	14,515	49,267	211.4
1976	127,980	13,311	54,934	213.0
1977	122,810	12,766	56,539	214.7
1978	116,375	12,348	60,100	216.5
1979	110,864	12,220	66,590	218.5
1980 1/	110,961	12,513		

1/ Preliminary

2/ Beginning in 1962 figures are Dec. 1 of previous year.

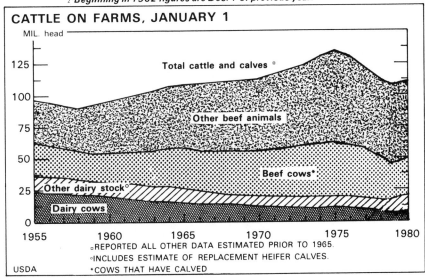

CATTLE ON FARMS, JANUARY 1

MIL. head

Total cattle and calves °

Other beef animals

Beef cows*

Other dairy stock°

Dairy cows

°REPORTED ALL OTHER DATA ESTIMATED PRIOR TO 1965.

°INCLUDES ESTIMATE OF REPLACEMENT HEIFER CALVES.

USDA *COWS THAT HAVE CALVED

Number of Cattle Feedlots and
Fed Cattle Marketed, 1978-79

	1978		1979	
State	Number Lots	Cattle Marketed (1,000 head)	Number Lots	Cattle Marketed (1,000 head)
Arizona	31	633	29	668
California	130	1,415	117	1,362
Colorado	420	2,451	390	2,239
Idaho	350	495	350	511
Illinois	14,000	980	13,500	920
Indiana	10,100	396	9,900	367
Iowa	33,000	3,242	32,000	2,890
Kansas	5,500	3,471	5,000	3,214
Michigan	1,500	271	1,500	219
Minnesota	10,900	755	10,900	700
Missouri	8,000	295	7,000	230
Montana	119	118	73	79
Nebraska	13,879	4,170	13,380	3,975
New Mexico	34	337	35	343
North Dakota	1,500	66	1,500	70
Ohio	7,000	382	6,300	300
Oklahoma	360	833	340	669
Oregon	500	175	500	143
Pennsylvania	6,000	110	6,000	104
South Dakota	8,400	555	8,000	575
Texas	1,120	4,915	1,050	4,445
Washington	82	406	72	406
Wisconsin	4,500	170	5,500	171
23 States	127,425	26,645	123,436	24,600

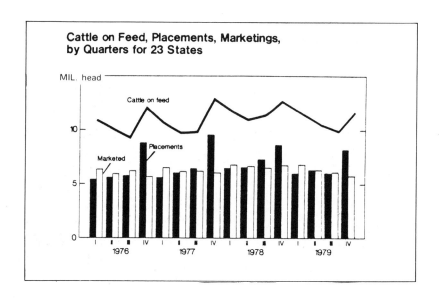

Cattle on Feed, Placements, Marketings, by Quarters for 23 States

Cattle on Feed, Placements & Marketings 1960-1979

UNITED STATES

(1,000 head)

Year	Number On Feed (Jan. 1)	Placements	Marketings
1960	7,064	13,525	13,060
1961	7,529	13,525	13,995
1962	7,919	15,749	14,560
1963	9,108	16,066	15,918
1964	9,256	17,491	17,366
1965	9,381	18,512	17,926
1966	9,967	20,256	19,534
1967	10,698	21,079	20,942
1968	10,835	23,792	22,662
1969	11,965	24,539	23,860
1970	12,644	24,449	24,884
1971	12,209	26,402	25,281
1972	13,330	27,376	26,845
1973	13,861	24,451	25,304
1974	13,067	19,886	23,334
1975	9,619	24,593	20,818
1976	12,296	25,464	24,151
1977	11,928	27,657	24,861
1978	12,809	29,073	26,645
1979	12,681	26,062	24,600
1980	11,739		

U. S. total equals the 23 major cattle feeding states.

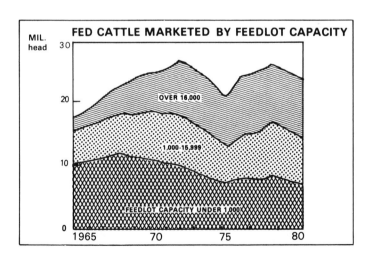

Total Meat Production and
Consumption in the United States, 1920 to Date

	BEEF		VEAL		PORK	
Year	Commercial Production MIL. Lbs.	Consumption Per Capita Lb.	Commercial Production MIL. Lbs.	Consumption Per Capita Lb.	Commercial Production MIL. Lbs.	Consumption Per Capita Lb.
1920	6,306	59.1	842	8.0	7,697	65.7
1925	6,878	59.5	989	8.6	8,128	66.8
1930	5,917	48.9	792	6.4	8,482	67.0
1935	6,608	53.2	1,023	8.5	5,919	48.4
1936	7,358	60.5	1,075	8.4	7,474	55.1
1937	6,798	55.2	1,108	8.6	6,951	55.8
1938	6,908	54.4	994	7.6	7,680	58.2
1939	7,011	54.7	991	7.6	8,660	64.7
1940	7,175	54.9	981	7.4	10,044	73.5
1941	8,082	60.9	1,036	7.6	9,528	68.4
1942	8,843	61.2	1,151	8.2	10,876	63.7
1943	8,571	53.3	1,167	8.2	13,640	78.9
1944	9,112	55.6	1,738	12.4	13,304	79.5
1945	10,276	59.4	1,664	11.9	10,697	66.6
1946	9,373	61.6	1,443	10.0	11,130	75.8
1947	10,432	69.6	1,605	10.8	10,502	69.6
1948	9,075	63.1	1,423	9.5	10,055	67.8
1949	9,439	63.9	1,334	8.9	10,286	67.7
1950	9,534	63.4	1,230	8.0	10,714	69.2
1951	8,837	56.1	1,059	6.6	11,481	71.9
1952	9,650	62.2	1,169	7.2	11,527	72.4
1953	12,407	77.6	1,546	9.5	10,006	63.5
1954	12,963	80.1	1,647	10.0	9,870	60.0
1955	13,569	82.0	1,578	9.4	12,295[2]	81.8
1956	14,462	85.4	1,632	9.5	12,675	82.7
1957	14,202	84.6	1,526	8.8	11,785	75.1
1958	13,330	80.5	1,186	6.7	11,658	73.0
1959	13,580	81.4	1,008	5.7	13,496	82.0
1960	14,753	85.1	1,109	6.1	13,026	77.6
1961	15,327	87.8	1,044	5.6	12,851	74.2
1962	15,324	88.9	1,015	5.5	13,258	75.0
1963	16,456	94.5	929	4.9	13,848	76.3
1964	18,456	99.9	1,013	5.2	14,033	76.2
1965	18,727	99.5	1,020	5.2	12,327	67.2
1966	19,726	104.2	910	4.6	12,576	65.7
1967	20,219	106.5	792	3.8	13,912	72.0
1968	20,880	109.7	734	3.6	14,104	72.4
1969	21,158	110.8	673	3.3	13,860	70.4
1970	21,684	113.7	588	2.9	14,500	72.7
1971	21,697	113.0	516	2.7	15,815	79.0
1972	22,218	116.1	429	2.2	14,241	71.3
1973	21,088	109.6	325	1.8	13,043	63.9
1974	22,844	116.8	442	2.3	14,100	69.1
1975	23,673	120.1	827	4.2	11,585	56.1
1976	25,667	129.3	813	4.0	12,488	59.6
1977	24,986	125.9	794	3.9	13,051	61.5
1978	24,010	120.0	600	3.0	13,209	61.5
1979	21,254	107.5	413	2.1	15,290	70.4

1/ Carcass Weight
2/ Series revised in 1977. Figures 1955 on reflect new series.

Meat Consumption Per Person

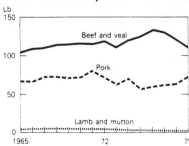

Commercial Slaughter of Cattle, Calves, Hogs, Sheep and Lambs
(1,000 head)

Year	Cattle	Calves	Hogs	Sheep and Lambs
1925	14,704	9,936	65,508	15,430
1930	12,056	7,761	67,272	21,125
1935	14,805	9,632	46,011	22,000
1940	14,958	9,089	77,610	21,571
1950	17,901	9,973	69,543	12,582
1951	16,376	8,418	76,061	11,075
1952	17,856	8,894	77,690	13,962
1953	23,605	11,668	66,913	15,967
1954	25,017	12,746	64,827	15,920
1955	25,722	12,377	74,216	16,215
1956	26,862	12,512	78,513	15,993
1957	26,232	11,904	72,595	14,957
1958	23,555	9,315	70,965	14,164
1959	22,930	7,683	81,582	15,180
1960	25,224	8,225	79,036	15,899
1961	25,635	7,701	77,335	17,190
1962	26,083	7,494	79,334	16,837
1963	27,232	6,833	83,324	15,822
1964	30,818	7,254	83,018	14,595
1965	32,347	7,420	73,784	13,006
1966	33,727	6,647	74,011	12,737
1967	33,869	5,919	82,124	12,791
1968	35,026	5,443	85,160	11,884
1969	35,237	4,863	83,838	10,691
1970	35,025	4,072	85,817	10,552
1971	35,585	3,689	94,438	10,729
1972	35,779	3,053	84,707	10,301
1973	33,687	2,249	76,795	9,597
1974	36,812	2,987	81,762	8,847
1975	40,911	5,209	68,687	7,835
1976	42,645	5,350	73,784	6,714
1977	41,856	5,517	77,303	6,355
1978	39,553	4,170	77,315	5,369
1979	33,650	2,824	89,089	5,017

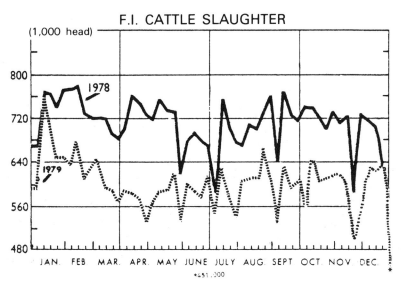

F.I. CATTLE SLAUGHTER

(1,000 head)

1978

1979

800

720

640

560

480

JAN. FEB. MAR. APR. MAY JUNE JULY AUG. SEPT. OCT. NOV. DEC.

*451.200

Retail Value of Consumption
Per Person of Beef and Pork, 1949-1979

	BEEF			PORK		
Year	Retail Weight	Average Retail Price Per Lb.	Retail Value Per Capita	Retail Weight	Average Retail Price Per Lb.	Retail Value Per Capita
1949	50.5	67.8	$ 34.24	63.0	54.5	$34.34
1950	50.1	74.6	37.37	64.4	53.8	34.65
1951	44.3	87.3	38.67	66.8	57.8	38.61
1952	49.1	85.7	42.08	67.4	56.2	37.88
1953	61.3	68.4	41.93	59.1	62.1	36.70
1954	62.9	67.8	42.65	55.8	63.4	35.38
1955	64.0	66.8	42.75	62.1	53.6	33.29
1956	66.2	65.4	43.29	62.5	51.4	32.13
1957	65.1	69.9	45.50	56.8	59.4	33.74
1958	61.6	80.2	49.40	56.0	63.8	35.73
1959	61.9	82.0	50.76	62.8	56.3	35.36
1960	64.2	80.2	51.49	60.3	55.9	33.71
1961	65.8	78.4	51.49	57.7	58.4	33.70
1962	66.2	81.7	54.09	59.1	58.8	34.75
1963	69.9	78.5	54.87	60.8	56.6	34.41
1964	73.9	76.5	56.53	60.8	55.9	33.99
1965	73.6	80.1	58.95	54.6	65.8	35.93
1966	77.1	82.4	63.53	54.0	74.0	39.96
1967	78.8	82.6	65.09	59.6	67.2	40.05
1968	81.2	86.6	70.32	61.6	67.4	41.52
1969	82.0	96.2	78.88	60.5	74.3	44.95
1970	84.1	101.7	85.46	62.0	77.4	48.00
1971	83.6	108.1	90.43	68.2	69.8	47.61
1972	85.9	118.7	101.92	62.9	82.7	52.01
1973	81.1	142.1	115.24	57.6	109.2	62.88
1974	86.4	146.3	126.41	62.2	107.8	67.11
1975	88.9	154.8	137.68	51.2	134.6	68.90
1976	95.6	148.2	141.84	54.6	134.0	73.20
1977	93.2	148.4	138.23	56.7	125.4	71.04
1978	88.8	181.9	161.52	56.6	143.6	81.25
1979*	79.6	226.3	180.02	64.8	144.1	93.33

*Preliminary
Note: Beef prices are for Choice grades; pork prices from selected cuts.
Source: U. S. Department of Agriculture.

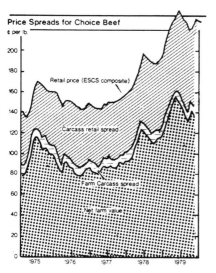

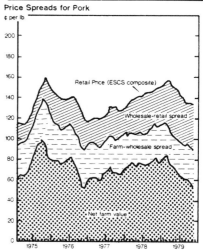

Per Capita Disposable Income and Expenditures for All Food, Beef and Pork, 1949-79

Based on revised retail prices and calculated retail
weight of consumption per person

Year	Disposable Income	EXPENDITURES			PERCENT OF DISPOSABLE INCOME		
		All Food	Beef	Pork	All Food	Beef	Pork
1949	$1,264	$300	$34.24	$34.34	23.8	2.7	2.7
1950	1,364	303	37.37	34.65	22.2	2.7	2.5
1951	1,468	338	38.67	38.61	23.0	2.6	2.6
1952	1,518	348	42.08	37.88	23.0	2.8	2.5
1953	1,582	347	41.93	36.70	22.0	2.7	2.3
1954	1,585	348	42.65	35.38	22.0	2.7	2.2
1955	1,654	351	42.75	33.29	21.1	2.6	2.0
1956	1,743	359	43.29	32.13	20.6	2.5	1.8
1957	1,801	373	45.50	33.74	20.7	2.5	1.9
1958	1,831	383	49.40	35.73	20.9	2.7	2.0
1959	1,905	386	50.76	35.36	20.3	2.7	1.9
1960	1,934	391	51.57	33.71	20.2	2.7	1.7
1961	1,984	397	51.59	33.70	20.0	2.6	1.7
1962	2,066	399	54.09	34.75	19.3	2.6	1.7
1963	2,139	402	54.87	34.41	18.8	2.6	1.6
1964	2,284	418	56.53	33.99	18.3	2.5	1.5
1965	2,430	443	58.95	35.93	18.2	2.4	1.5
1966	2,605	472	63.53	39.96	18.1	2.4	1.5
1967	2,749	481	65.09	40.05	17.5	2.4	1.5
1968	2,946	516	70.32	41.52	17.5	2.4	1.4
1969	3,130	541	78.88	44.95	17.3	2.5	1.4
1970	3,348	584	82.92	40.38	17.3	2.5	1.4
1971	3,588	591	87.21	47.95	16.4	2.4	1.3
1972	3,837	626	97.75	52.33	16.3	2.5	1.4
1973	4,285	700	109.88	63.20	16.3	2.6	1.5
1974	4,646	790	119.91	67.36	17.0	2.6	1.4
1975	5,077	863	129.79	69.10	17.0	2.5	1.4
1976	5,511	926	132.84	72.86	16.8	2.4	1.3
1977	6,035	1,002	127.77	71.39	16.6	2.1	1.2
1978	6,667	1,106	162.08	81.57	16.6	2.4	1.2
1979*	7,362	1,220	180.27	93.94	16.6	2.4	1.3

Preliminary

INCOME AND EXPENDITURES

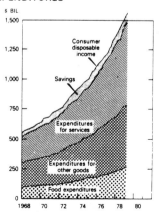

Consumer disposable income does not include interest paid by consumers and personal transfer payments to foreigners.
Source: U.S. Department of Commerce, seasonally adjusted annual rates.

U. S. Imports, Exports and Net Imports of Beef, Veal and Total Meat in Relation to Domestic Production 1970-1979

BEEF AND VEAL

Year	Production*	Imports*	Exports*	Net Imports*	Percentage of U. S. Production		
					Imports	Exports	Net Imports
1970	9,992,742	612,415	13,291	599,124	6.1	.1	6.0
1971	10,075,750	594,543	19,052	575,491	5.9	.2	5.7
1972	10,272,611	671,746	23,678	648,068	6.3	.2	6.1
1973	9,712,873	678,913	35,880	643,033	7.0	.4	6.6
1974	10,562,460	552,448	22,953	529,496	5.2	.2	5.0
1975	11,113,127	596,357	20,694	575,663	5.4	.2	5.2
1976	12,011,249	672,284	36,555	635,729	5.6	.3	5.3
1977	11,694,184	633,636	41,498	592,138	5.4	.4	5.0
1978	11,160,998	750,883	55,020	695,863	6.7	.5	6.2
1979	9,826,304	787,969	57,403	730,562	8.0	.6	7.4

TOTAL MEATS

Year	Production*	Imports*	Exports*	Net Imports*	Imports	Exports	Net Imports
1970	16,246,938	820,892	48,853	772,039	5.0	.3	4.7
1971	16,947,745	798,348	56,701	741,647	4.7	.3	4.4
1972	16,619,794	902,632	76,569	826,063	5.4	.5	4.9
1973	15,646,829	886,212	119,299	766,913	5.7	.8	4.9
1974	16,929,601	733,573	78,383	655,190	4.3	.5	3.8
1975	16,426,109	762,838	121,050	641,788	4.6	.7	3.9
1976	17,717,046	838,375	185,644	652,731	4.7	1.0	3.7
1977	17,768,756	783,979	180,047	603,932	4.4	1.0	3.4
1978	17,287,528	931,726	167,016	764,710	5.4	1.0	4.4
1979	16,888,889	978,942	163,016	815,926	5.8	1.0	4.8

*Metric tons

Beef and Veal Imports and Exports by Countries, 1978-1979

(Metric Tons)

Beef and Veal Imported From:

	1978	1979
Argentina	48,377	51,482
Australia	368,222	397,701
Brazil	18,820	17,767
Canada	28,735	35,268
Costa Rica	28,443	32,233
Denmark	805	1,047
Dominican Republic	812	1,717
El Salvador	3,833	4,888
Guatemala	13,444	15,561
Honduras	19,485	28,090
Ireland	4	19
Mexico	28,733	2,424
New Zealand	153,886	161,562
Nicaragua	31,103	33,822
Paraguay	3,613	886
Uruguay	1,136	1,698
Other countries	1,432	1,800
Total	750,883	787,965

Beef and Veal Exported To:

	1978	1979
Canada	4,049	3,983
Caribbean area	6,626	6,807
European countries (9)	1,395	1,066
Hong Kong	581	625
Japan	33,845	35,254
Mexico	765	775
Saudi Arabia	1,974	2,281
Taiwan	283	2,250
Other countries	5,502	4,362
Total	55,020	57,403

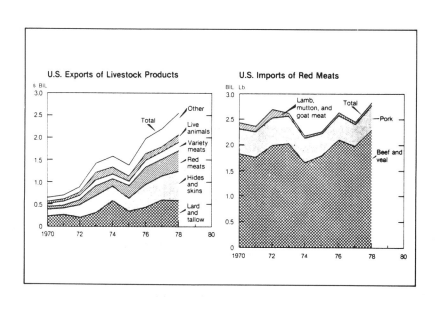

Feed Grains (Total)
Acreage, Supply and Distribution, 1974-80
FEED GRAINS (CORN, SORGHUM, OATS AND BARLEY)

	1974/75	1975-76	1976/77	1977/78	1978/79	1979/80
					Prel.	Est. 1/
ACREAGE (MIL.)						
Base or allotment	89.0	89.0	89.0	89.0	97.4	106.7
Set-aside	—	—	—	—	8.3	4.6
Planted	122.5	123.3	128.7	128.9	122.8	117.6
Harvested for grain	100.6	105.1	106.3	108.1	104.5	101.2
METRIC TONS						
Yield per acre	1.49	1.75	1.82	1.88	2.08	2.31
MILLION METRIC TONS						
Supply						
Beginning stocks 2/	21.5	15.2	17.2	29.9	41.2	45.9
Production	150.0	184.4	193.4	203.4	217.4	233.9
Imports	.5	.5	.4	.3	.3	.2
Total	172.0	200.1	211.0	233.6	258.9	280.0
Disappearance						
Feed	105.0	115.8	112.6	117.3	133.1	136.9
Food, industry & seed	16.1	17.1	17.9	18.8	19.7	20.7
Total domestic	121.1	132.9	130.5	136.1	152.8	157.6
Exports	35.7	50.0	50.6	56.3	60.2	65.9
Total	156.8	182.9	181.1	192.4	213.0	223.5
Ending stocks						
Government 3/	.1	0	0	10.8	20.2	29.1
Free 4/	15.1	17.2	29.9	30.4	25.6	27.4
Total	15.2	17.2	29.9	41.2	45.8	56.5

1/Based on January indications
2/October 1 corn and sorghum; July 1 oats and barley
3/Under loan and uncommitted CCC inventory
4/Privately owned stocks

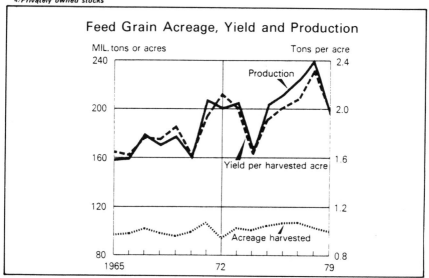
Feed Grain Acreage, Yield and Production

LITERATURE CITED

ABS. 1979. Androgenized cows. American Breeder's Service, Vol.II no.3.

Albert, W.W. 1969. Good performance from beef cows confined to drylot the year round. Ill. Res. Univ. of Ill. Agr. Exp. Sta. Bull.

Beverly, J.R. 1975. Recognizing and handling calving problems. Tex. Agr. Ext. Ser. MP-1203.

Beverly, J.R. 1978. The outlook for estrous synchronization. Presented West. Sec. ASAS. Cal Poly, Pomona, Calif. June 14-16.

Breimyer, H.F., J. Tucker, H.F. Carman, R.S. Fenwick and W.F. Woods. 1975. Who will control U.S. agriculture? North Cent. Reg. Pub. #37, Univ. of Ill.

Brinks, J.S. and B.W. Knapp. 1975. Effects of inbreeding on performance traits of beef cattle in the western region. Colo. State Univ. Exp. Sta. Tech. Bull. 123.

Byers, F.M., J.K. Matsushima and D.E. Johnson. 1975. Application of the concept of associative effects of feeds to prediction of ingredient and diet energy values. Beef Nut. Res. Colo. State Univ. Exp. Sta. Gen. Series 947:22.

Church, D.C. 1969. Digestive Physiology and Nutrition of Ruminants, Vol. I - Digestive Physiology. O&B Books, Corvallis, Ore.

Church, D.C. 1974. Digestive Physiology and Nutrition of Ruminants Vol. 2 - Nutrition. O&B Books, Corvallis, Ore.

Church, D.C., L.S. Pope and W.A. McVicar. 1956. Effect of plane of nutrition of beef cows upon depletion of liver vit. A during gestation and on carotene requirements during lactation. J. Anim. Sci. 15:1078.

Coats, W.L. 1974. The principles of tax reform. Challenge. Jan.-Feb.

Corah, L.R., T.G. Dunn, C.C. Kaltenbach and C.O. Schoonover. 1974. Univ. of Wyo. Agr. Ext. Ser. Bull. #616.

CME. 1978. Cattle Hedger's Handbook. Chicago Mercantile Exchange.

Dam, A. 1972. Immunoprophylaxis in colibacillosis in calves, especially by treatment with specific E. coli antiserum. Acta. Vet. Brno. Supp. 2:87.

Davenport, C. 1975. Income tax law and regulations affecting agriculture. Spec. Rpt. #172. Agr. Exp. Sta. Univ. of Mo.

Davis, G.K. and H.F. Roberts. 1959. Urea toxicity in cattle. Fla. Agr. Exp. Sta. Bull. 611.

Davis, G.V. and O.T. Stallcup. 1967. J. Dairy Sci. 50:1638. Adapted from: D.C. Church. 1969.

Deutscher, G.J., A.L. Slyter and L. Blone. 1975. Effects of mating and management systems on performance of beef cows and calves. So. Dakota Agr. Exp. Sta. Anim. Sci. Series 75-2.

DHEW. 1971. Arteriosclerosis Vol. II. Dept. Health Educ. & Welfare Pub. no. (NIH) 72-219.

Eisgruber, L. 1976. Dept. Head, Dept. Agr. Econ., Ore. State Univ. Sem.: "Economic analysis of grazing lands." Personal communication.

Ewing, S.A., W. Burroughs, W. Woods and C.C. Culbertson. 1959. Iowa State Coll. Anim. Husb. Leaflet #427.

Gallup, W.D., L.S. Pope and C.K. Whitehair. 1953. Urea in rations for cattle and sheep. Okla. Agr. Exp. Sta. Bull. 409.

Gray, E.F., F.A. Thrift and C.W. Absher. 1978. Heterosis expression for preweaning traits under commerical beef cattle conditions. J. Anim. Sci. 47:370.

Groen, J., B.K. Tjong, C.E. Kamminga, et al. 1952. The influence of nutrition, individuality and some other factors, including various forms of stress, on the serum cholesterol; an experiment of nine months duration in 60 normal human volunteers. Voeding 13:556-587. Adapted from: DHEW. 1971.

Hammack, S.P., P.T. Marion and J.K. Riggs. 1969. Semi-confinement production of crossbred calves for beef. Tex. Agr. Exp. Sta. Cons. PR - 2676-2695.

Harris, R.R., V.L. Brown, W.B. Anthony and C.C. King. 1970. Confined feeding of brood cows. Auburn Univ. Exp. Sta. Bull. #411.

Harrison, V.L. 1973. Accounting methods allowed farmers: tax incentives and consequences. USDA Econ. Res. Ser. Bull. #505.

Ikerd, J.E. 1977. Managing total market risk in feeding operations. Okla. 13th Annual Cattle Feeder's Sem. Okla. State Univ.

Kahn, H.A., J.H. Medaliw, H.N. Neufeld, et al. 1969. Serum cholesterol: its distribution and association with dietary and other variables in a survey of 10,000 men. Israel J. Med. Sci. 5:1117-1127. Adapted from: DHEW. 1971.

Kannel, W.B. and T. Gordon (ed.s). The Framingham Study: an epidemiological investigation of cardiovascular diseases, section 24, diet and the regulation of serum cholesterol. U.S. Dept. of Health, Educ. and Welfare, Pub. Health Ser., Nat. Inst. of Health. Adapted from: DHEW. 1971.

Kieffer, N.M. and T.C. Cartwright. 1967. Domestic cattle chromosomes In Vitro. I. Analysis of diploid number and morphology. Tex. Agr. Exp. Sta. Cons. PR - 2483-2500.

Long, C.R. 1980. Crossbreeding for beef production: experimental results. J. Anim. Sci. 51:1197.

Lusby, K.S. and S.L. Armbruster. 1976. Winter supplementation - not substitution. Okla. State Univ. Anim. Sci. Ext. Ser. Misc. Publ.

Matsushima, J.K. and W.E. Smith. 1975. Synovex implant - once or twice? Beef Nut. Res. Colo. State Univ. Exp. Sta. Gen. Series 947:12.

McCartor, M.M. 1972. Effects of dietary energy and early weaning on the reproductive performance of first-calf Hereford heifers. Tex. Agr. Exp. Sta. Cons. PR - 3111-3131.

McCloskey, W.B. Jr. 1979. Massed fish. Oceans. Vol. 12 no. 1.

McGinty, D.D. and M. Belew. 1972. Drylot cow-calf production. Univ. of Ariz. Anim. Sci. Misc. Bull.

McPeake, C.A. and R.R. Frahm. 1980. The genetics of crossbreeding. Okla. Beef Symp. Okla. State. Univ.

Meisner, J.C. and V.J. Rhodes. 1974. The changing structure of U.S. cattle feeding. Spec. Rpt. #167. Univ. of Mo.

Melton, A.A., T.C. Cartwright and W.E. Kruse. 1967. Cow size as related to efficiency of beef production. Tex. Agr. Exp. Sta. PR - 2483-2500.

Melton, A.A. and J.K. Riggs. 1964. Frequency of feeding protein supplement to range cattle. Tex. Agr. Exp. Sta. B-1025.

Morrison, F.B. 1961. Feeds and Feeding Abridged. Morrison Publ. Co. Clinton, Iowa.

Oppenheimer, H.L. 1972. Cowboy Litigation - Cattle and the Income Tax. Interstate Publ., Danville, Ill.

Perry, T.W., R.C. Peterson and W.M. Beeson. 1974. A comparison of drylot and conventional herd management systems. J. Anim. Sci. 38:249.

Pollak, E.J. and G.R. Ufford. 1978. Effect of inbreeding on within-herd genetic evaluation of beef cattle. J. Anim. Sci. 47:853.

Preston, T.R. and M.B. Willis. 1970. Intensive Beef Production. 2nd ed. Pergamon Press, Oxford.

Price, D.P. 1977. Cow-calf confinement energy levels. Ph.D. Dissertation, Ore. State Univ., Corvallis.

Price, D.P. and A.T. Ralston. 1975. Confinement energy levels for beef cows. Agr. Exp. Sta. Spec. Rpt. #437. Ore. State Univ.

Price, T.D. 1973. Percent dystocia in Angus and Angus x Hereford heifers. Colo. State Univ. Res. Highlights.

Price, T.D. and J.N. Wiltbank. 1978. Dystocia in cattle a review and implications. Theriogenology, Vol. 9 no. 3.

Prigge, E.C., M.L. Galyean, F.N. Owens, D.G. Wagner and R.R. Johnson. 1978. Microbial protein synthesis in steers fed processed corn rations. J. Anim. Sci. 46:249.

Purcell, W.D. 1977. Managing total market risk in feeding operations: empirical applications. Okla. 13th Annual Cattle Feeder's Sem. Okla. State Univ.

Radostits, O.M., C.S. Rhodes, M.E. Mitchell, T.P. Spotsworth and M.S. Wenkoff. 1975. A clinical evaluation of antimicrobial agents and temporary starvation in the treatment of acute undifferentiated diarrhea in newborn calves. Can. Vet. J. 16:219.

Ragsdale, B.J., D.L. Huss and G.O. Hoffman. 1971. Grazing systems for profitable ranching. Tex. Agr. Ext. Ser. MP-896.

Reynolds. 1953. Adapted from: J.K. Riggs. Fifty years of beef cattle nutrition. J. Anim. Sci. Vol. 17, no. 4.

Rhodes, V.S. 1973. Consequences of income tax law and regulations: cattle feeding. Agr. Exp. Sta. Bull. #172. Univ. of Mo.

Rice, V.A., F.N. Andrews, E.J. Warick and J.E. Legates. 1970. Breeding and Improvement of Farm Animals. McGraw-Hill Book Co., New York.

Riggs, J.K. and P.T. Marion. 1972. Maintaining beef cows in confinement. Chap. 15. In: O'Mary, C.C. and I.A. Dyer. Commercial Beef Cattle Production, p. 248. Lea & Febiger, Philadelphia.

Ritchie, H.D., H.D. Woody, C.A. McPeake and W.T. Magee. 1975. Third-year calving and weaning performance of cows maintained in drylot year-round. Mich. State Univ. Anim. Husb. Rpt. 7407.

Ritchie, H.D., H.D. Woody, D.P. Olson, H.E. Henderson and D.R. Strohbehn. 1973. First-year calving performance of beef cows maintained in drylot year-round. Mich. State Univ. Anim. Husb. Rpt. 7203.

Ruttle, J. 1980. Fertility of New Mexico range bulls. New Mexico Stockman, Dec.

Simpson, F.H. 1974. Tax rule changes could empty pens. Beef Magazine. Jan.

Slyter, A.L. 1970. Influence of mating and management systems on the performance of beef cows and calves. So. Dakota Beef Cattle Field Day.

Tanaka, N. and O.W. Portman. 1977. Effect of type of dietary fat and cholesterol on cholesterol absorption rate in Squirrel Monkeys. Ore. Reg. Primate Res. Cntr. Pub. #884.

Taylor, B., S. Swingle, M. Selke, A. Phillips and C. Roe. 1975. Low quality roughages for cows nursing calves. Univ. of Ariz. Cattle Feeder's Day. Series P-36.

Thomas, R.C. and T.C. Cartwright. 1971. Efficiency of F1 Angus-Jersey cows in three-way crosses. Tex. Agr. Exp. Sta. Cons. PR - 2963-2999.

Totusek, R. 1975. Beef cow efficiency as affected by size and milk production. Ore. Agr. Exp. Spec. Rpt. #437.

USDA. 1975. Controlling grass tetany. Leaflet no. 561.

Wagner, D.G., R.P. Wettemann and J.C. Aimone. 1976. Reimplanting studies with feedlot cattle. Okla. Agr. Exp. Sta. Anim. Sci. Res.

Wallace, J.D. and R.J. Raleigh. 1964. Calf production from Hereford cows wintered at different nutritional levels. Proc. West. Sec. Amer. Soc. Anim. Sci. 15:57.

Willet, G.S., E.L. Mezie and A.G. Nelson. 1974. Cattle feeding tax benefits to non-farm investors. Univ. of Ariz. and ERS, USDA.

Wiltbank, J.N. 1958. Report of beef cattle research activities. Univ. of Neb. From: J.K. Riggs, Nutrition requirements of beef females for maximum production. Tex. A&M Ext. Pub.

Wiltbank, J.N. 1966. Influence of total feed and protein intake on reproductive performance in beef heifers as affected by protein and energy intake during gestation. J. Anim. Sci. 20:957. (Abstract)

Wiltbank, J.N. 1972. Management of heifer replacements and the brood cow herd through the calving and breeding periods. Chap. 11. In: O'Mary, C.C. and I.A. Dyer. Commercial Beef Cattle Production, p. 150. Lea & Febiger, Philadelphia.

Wiltbank, J.N., W.W. Rowden, J.E. Ingalls, K.E. Gregory and R.M. Coch. 1965. Effect of energy level upon reproductive phenomena of Hereford cows. J. Anim. Sci. 21:219.

Woods, W.F. and T.A. Carlin. 1975. Utilization of special farm tax rules. Agr. Exp. Sta. Spec. Rpt. #172. Univ. of Mo.

Wooten, R.A., C.B. Roubicek, J.A. Marchello, F.D. Dryden and R.S. Swingle. 1979. Realimentation of cull range cows. 2. changes in carcass traits. J. Anim. Sci. 48:823.

Yudkin, J. 1968. Dietary intake of carbohydrate in relation to diabetes and atherosclerosis. Chap. 7. In: Dickens, F., P.J. Randle and W.J. Whelan. Carbohydrate Metabolism and its Disorders, p. 169. Academic Press, London.

INDEX

348

GLOSSARY
(If term is not listed in glossary, check the index.)

amino acid - Organic molecules containing nitrogen necessary for the formation of protein; e.g. the building blocks of protein. A feed is said to be "balanced" for amino acids if it contains all the amino acids the animal requires for the formation of body tissue.

antibodies - Substances formed within the body of an animal for combatting disease organisms.

Bangs - Brucellosis. Banger - a cow that gives a positive reaction to a Brucellosis test.

buffer - A compound used to minimize the acidity or alkalinity of a fluid.

calcification - the deposition of calcium in soft tissue.

capital gains - Special tax treatment given to income derived from the sale of an item ordinarily used for the production of income; e.g. manufacturing equipment, common stock, etc. The tax liability is less than charged for"ordinary income"; the purpose being to stimulate investment and thereby stimulate the economy.

carcinogen - A compound that tends to cause cancer.

carotene - Compounds found in plants that can be used by animals such as ruminants, for synthesis of vit. A.

cholesterol - An organic compound used in the formation of body hormones and other compounds. Cholesterol is synthesized in the body and can be absorbed from the gut as well. It is implicated in blockage of the blood vessels (arteriosclerosis) and subsequent heart disease. Although intensly studied, much of cholesterol metabolism remains to be discovered.

clinical - As a term used to describe disease, the word clinical means the animal is noticeably ill. This is opposed to the term subclinical, which means the animal contracts the disease, but outwardly appears to be healthy.

coccidiostat - A compound used or capable of combatting coccidiosis, a very common parasitic disease of the gut.

cod fat - The fat contained in the crotch of an animal.

clostridial diseases - Diseases caused by the genera of bacteria known as Clostridia. These bacteria create a powerfull toxin which can cause rapid and sudden death losses; e.g. Blackleg, Malignant Edema, etc.

colostrum - The first milk given by a mammal after parturition. This milk is much higher in protein, fat, and sugars than ordinary milk, and contains antibodies. In the case of cattle, it is essential that the newborn calf receive colostrum, as antibodies cannot be passed accross the bovine uterine wall. Without colostrum, the calf has no immunity against disease.

concentrate - A feedstuff high in energy and/or protein; e.g. grains and oilmeals.

dermatitis - A skin disorder which may result in flakiness, pustules, or otherwise obviously unhealthy appearance.

doggie - orphaned calf.

double muscling (doppellender disease) - A condition that has occured in lines of cattle bred for extreme muscling. Affected cattle have extra muscle fibers, and outwardly exhibit an extremely muscular appearance. The situation usually results in reproductive problems.

encephalitic condition - A disorder that results in neurological dysfunction.

enzymes - Compounds that are necessary or otherwise facilitate biochemical reactions; e.g. proteolytic enzymes cause the breakdown of dietary proteins into their constituent amino acids for absorption.

equity - Commonly referred to as the amount of money an investor has placed in an investment that has come out of his own pocket; i.e. contrasted to the amount of borrowed money; e.g. a 30% equity investment means the investor has put up 30% of the money and borrowed 70%.

fibrilate - A term describing a dysfunction of the heart. A fibriating heart is one that does not beat with its normal rhythm, but rather, spasmotically contracts and relaxes at an increased, but irregular and disorganized pace. The end result is that blood is not pumped through the circulatory system, and if not corrected, will usually result in death.

glut - A term used to describe the condition whereby an extraordinary amount of goods or commodity are placed on the market, thereby driving down the price.

glycogen - The form glucose (blood sugar) is stored in. When immediate energy is required, glycogen is released.

gonadotrophic hormones - hormones used to stimulate or control the reproductive functions and the expression of sexual characteristics.

grade (quality) - The quality grade in meat generally refers to the amount of marbling contained in the meat. The greater the marbling, the higher the grade. For slaughter steers and heifers the grades are (in descending order), Prime, Choice, Good, Standard, and Utility.

grade (yield) - Yield grade refers to the amount of fat contained on a carcass. The greater the fat the larger the yield grade number. The numbers run from #1 (leanest) to #5 (fattest).

hedging - Commonly referred to as using the future's market to guarantee a sale price and thereby reduce market risk.

heiferette - A feedlot term used to describe females that are too old to be called heifers, but are not old enough to display the angular characteristics of a cow. When fat, heiferettes will bring a price higher than cows, but not quite as high as heifers.

hemoglobin - The compound in the red blood cells used to carry and transport oxygen.

homeostasis - Commonly referred to as the net result of a myriad of biochemical mechanisms which are used to keep the content of the blood uniform; i.e. glucose, pH, etc.

jaundice - A yellowish tinge to the skin caused by increased bile in the blood.

legume - A type of plant that is capable of utilizing nitrogen from the air; i.e. does not require nitrogen fertilization. Indeed, legumes typically deposit nitrogen in the soil which can be used by other plants.

leppie - orphaned calf.

lipoproteins - Proteins complexed with fat (normally carried in the blood). Currently lipoproteins are under investigation as to a possible role in arteriosclerosis.

maintenance requirement - The amount of energy required by an animal to maintain its bodyweight.

marbling - Fat deposited within the muscle tissue.

maverick - An animal that has escaped its owner and has been foraging on its own (unmolested).

"Okie" cattle - Crossbred commercial cattle of mixed breeding.

ovarian cycle - The regular 21 day cycle whereby non-pregnant cows will go into extrus (heat), ovulate (release and egg cell), and then become sexually unreceptive until the next onset of estrus, 18 to 20 days later.

pathogen - A microorganism capable of causing disease.

pH - A measure of the acidity of alkalinity of a liquid. Measured on a logarithmic scale of 1 to 14, 7 is neutral, 1 is the most acid and 14 is the most alkaline.

photosynthesis - The process by which plants take sunlight and convert it into organic energy.

polled - The abscence of horns in cattle.

precursor - In biology, commonly referred to as a compound which may be synthesized by the animal into another compound, usually a vitamin.

saturated fats - Fats which have no double bonds between carbons. That is, fats that contain a full complement of hydrogen atoms. Saturated fats are usually hard at room temperature. Most animal fats are saturated, or "hard" fats.

stag - A steer that exhibits the physical characteristics of a bull. Usually an animal castrated after reaching sexual maturity.

steroids - Hormone and hormone-like substances, usually affecting sex and reproduction.

terminal sire - Bull used to sire calves for slaughter only; i.e. none of the calves are held for breeding purposes.

tetany - A condition involving powerful muscle spasms.

triglycerides - A saturated form of fat carried in the blood thought to associated with arterioschlerosis.

unsaturated fats - Fats in which two or more carbons will have double bonds (carry less that a full complement of hydrogen). Unsaturated fats are usually liquid at room temperature and are thus called "soft" fats. Most vegetable fats are unsaturated.

354

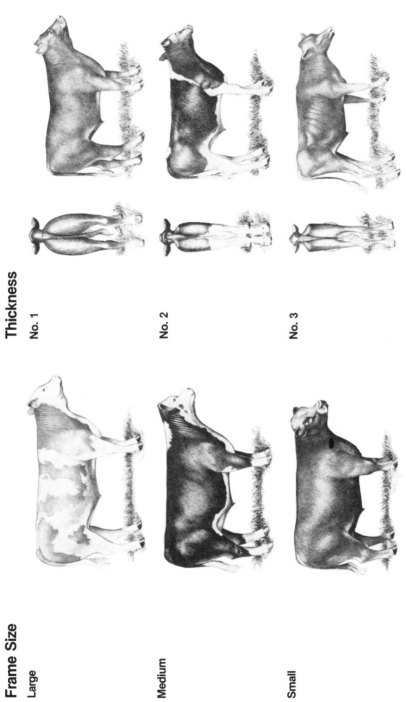

Frame Size

Large

Medium

Small

Thickness

No. 1

No. 2

No. 3

Large and medium frame pictures depict minimum grade requirements. The small frame picture represents an animal typical of the grade.

No. 1 and No. 2 thickness pictures depict minimum grade requirements. The No. 3 picture represents an animal typical of the grade.

RELATIONSHIP BETWEEN MARBLING, MATURITY, AND CARCASS QUALITY GRADE •

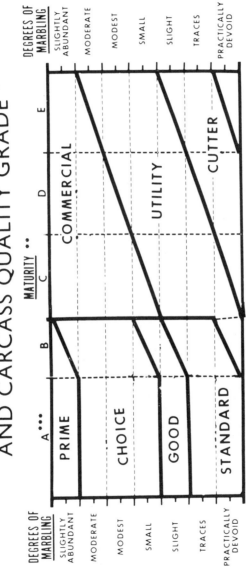

•Assumes that firmness of lean is comparably developed with the degree of marbling and that the carcass is not a "dark cutter."

••Maturity increases from left to right (A through E).

•••The A maturity portion of the Figure is the only portion applicable to bullock carcasses.

Figure 1

ILLUSTRATIONS OF THE LOWER LIMITS OF CERTAIN DEGREES OF TYPICAL MARBLING REFERRED TO IN THE OFFICIAL UNITED STATES STANDARDS FOR GRADES OF CARCASS BEEF

Illustrations adapted from negatives furnished by New York State College of Agriculture, Cornell University

1—Very abundant	4—Slightly abundant	7—Small
2—Abundant	5—Moderate	8—Slight
3—Moderately abundant	6—Modest	9—Traces

(Practically devoid not shown)

UNITED STATES DEPARTMENT OF AGRICULTURE
CONSUMER AND MARKETING SERVICE
LIVESTOCK DIVISION

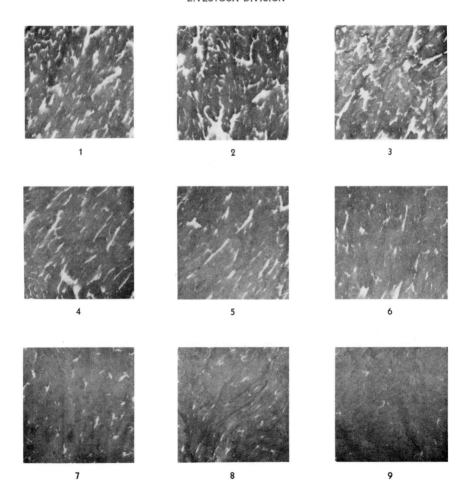

1

2

3

4

5

6

7

8

9

SLAUGHTER STEERS
U.S. GRADES
(YIELD)

YIELD GRADE 1 ——

YIELD GRADE 2 ——

YIELD GRADE 3 ——

YIELD GRADE 4 ——

YIELD GRADE 5 ——

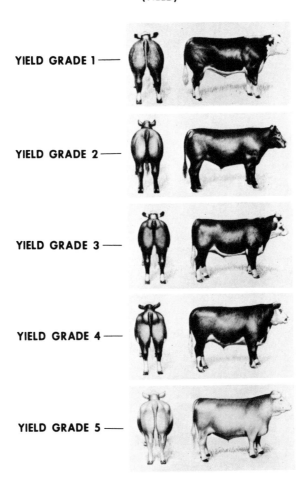

COPIES OF THE OFFICIAL
UNITED STATES STANDARDS
FOR GRADE ARE AVAILABLE
ON REQUEST

UNITED STATES DEPARTMENT OF AGRICULTURE
CONSUMER AND MARKETING SERVICE
LIVESTOCK DIVISION
WASHINGTON, D.C.

AUGUST 1969

Order Form

SWI Publishing
P.O. Drawer 3 A&M
University Park, NM 88003
U.S.A.
505-525-1370

_____ copies **Modern Agriculture** @ $34.50 _____
science, finance, production & economics

_____ copies **Beef Production** @ $24.95 _____
science & economics; application & reality
hardcover @ $32.00 _____

_____ copies **Intelligent Dieting** @ $13.95 _____
hardcover @ $17.95 _____
(A review of animal food products in
human nutrition.)

No shipping or handling charges.

20% discount on multiple books.

30 day unconditional guarantee on all books —
return for a full refund.

Name _____

Address _____

City _____

State, Zip _____

checks accepted or
VISA **MasterCard** VISA / MasterCard

Card No. _____

Expiration date _____

Signature _____